NIELS BOHR'S COMPLEMENTARITY

BOSTON STUDIES IN THE PHILOSOPHY OF SCIENCE

VOLUME 286

For further volumes:
http://www.springer.com/series/5710

NIELS BOHR'S COMPLEMENTARITY

Its Structure, History, and Intersections with Hermeneutics and Deconstruction

by

MAKOTO KATSUMORI

Akita University, Akita, Japan

Springer

Makoto Katsumori
Akita University
Faculty of Education and Human Studies
Tegata-Gakuen-machi 1-1
010-8502 Akita
Japan
katsumor@ipc.akita-u.ac.jp

ISSN 0068-0346
ISBN 978-94-007-1747-3 e-ISBN 978-94-007-1748-0
DOI 10.1007/978-94-007-1748-0
Springer Dordrecht Heidelberg London New York

Library of Congress Control Number: 2011931421

Printed on acid-free paper

Springer is part of Springer Science+Business Media (www.springer.com)

Preface

Niels Bohr, his words and his ideas, have occupied a considerable part of my mind ever since the early 1980s, when I began studying the history and philosophy of science. Bohr's discourse first sounded just puzzling and enigmatic, but I gradually became aware of its significance for the issues I was vitally concerned with – particularly a problematic that came up during the preceding years of my studies in physical science.

Up until my mid-twenties, as a student of geophysics, specifically meteorology, I was both intrigued and troubled by what seemed to be a large gap between nature as the object of scientific research and nature as experienced in other contexts of our life. This gap even appeared to render the use of the same word 'nature' questionable. The movement of air as described by the hydrodynamic equation of motion, for example, scarcely resembled the gentle rustling breeze I felt while walking in the woods or the raging storm whose overwhelming power inspired me with awe. The more I advanced in my research in meteorology, the more severely I felt this gap between the two notions of nature, or between our two different modes of relating to nature. In those days, I still naively held the belief that nature as described or explained scientifically is nature as it truly is, while most other aspects of nature disclosed in our life are *merely* subjective appearances thereof. At the same time, however, I could no longer suppress my acute feeling of losing contact with nature as I tried to approach the natural world along the lines of physical science. It was in order to engage squarely with this problem, to make it a theme of my own investigation, that I switched to the field of the history and philosophy of science.

At this juncture, I was struck by Bohr's idea of "complementarity," especially as expressed by the dictum that 'we are both spectators and actors in the great drama of existence.' Bohr discussed the relation of complementarity in quantum theory as well as other fields of knowledge and experience – the relation between space-time coordination and the claim of causality (corresponding, respectively, to the roles of the 'actor' and the 'spectator'), between psychical experience and its reflective analysis, and so on. It was crucial that, in his view, neither of the two complementary relata, neither of the roles of 'spectator' and 'actor,' has priority over the other. As I saw it, the implications of Bohr's complementarity extend so far as to bear essentially on the above problem I was wrestling with. Specifically, there appeared to be a sense in which the 'spectator' is the one who treats nature as the object of physical-scientific analysis, whereas the 'actor' experiences nature in a non-objectifying manner. At the same time, however, as emphasized by Bohr, the scientific, objectifying approach to nature also *acts* upon natural processes in its irreducible effect of observational intervention. Further, even the conceptual and discursive dimension of science might, in a sense, be regarded as being an 'actor,' if we can no longer simply separate nature in itself and our conception of nature. This suggested that the relation between the objectifying and the non-objectifying approaches to nature – or, more generally, the relation between the roles of 'spectators' and 'actors' – may be much more complex than they first appeared.

It was along this line of thought that I set out to study Bohr's work with a focus on complementarity, and, in due course, made it one of the main themes of my research. My approach slowly took shape as a combination of the following two lines of inquiry: On the one hand, I took pains with an intensive reading of Bohr's texts and a conceptual analysis thereof, especially of those elements of his thought which first seemed too subtle and elusive to grasp. On the other, I strove to extend my scope to a broader philosophical context in which to situate his idea of complementarity. Specifically, I gradually became convinced that his thought may be meaningfully compared and connected, not so much with approaches belonging to the analytic tradition of philosophy, but rather with such (so-called Continental) currents of thought as hermeneutic philosophy and Derridean deconstruction. I confess, however, that, during the early stages of my research, I often felt as if I alone had been walking a narrow and obscure path, with no one else understanding, let alone endorsing, my unconventional approach to Bohr. Yet, under these circumstances, it was all the more encouraging that, in the mid-nineties, I came across Arkady Plotnitsky's analysis of complementarity in its intersections with deconstruction, and later Karen Barad's and other authors' no less unorthodox approaches. I was again greatly encouraged when, at the 2003 Conference of the International Society for Hermeneutics and Science held in Tihany, Hungary, I met a number of researchers who took interest in my approach to complementarity – notably Hans Radder, who would soon kindly assume the role of my doctoral supervisor.

As it turned out, however, it took much more time and greater effort than initially expected to carry through my research project. While writing some separate articles on the subject, I spent a large part of my endeavor to weave together the different threads of my analysis of Bohr's thought into a more or less coherent texture. It was not until 2005 that I largely achieved this goal, when I completed my doctoral project at the Vrije Universiteit, the Netherlands, with a dissertation entitled "Niels Bohr's Complementarity: Its Structure, History, and Intersections with Hermeneutics and Deconstruction." The present book is based on this thesis, although it has been revised and expanded at a number of points.

Some ideas and arguments in the book have already appeared in preliminary forms elsewhere. In addition to the basic exposition of Bohr's complementarity in Chapter 2, the first half of Chapter 4 has, with substantial changes, grown out of my article "Bohr's Early Complementarity Argument," *Historia Scientiarum*, **8**,

No. 1 (1998), 1–19. The contents of other parts of Chapter 4 are given in outline in "Niels Bohr's Complementarity: A Philosophical Analysis," *Philosophia-Iwate*, No. 37 (2005), 34–51. Some major points in Chapters 5 and 6 have been provisionally discussed in "L'idée de complémentarité de Bohr en comparaison avec l'herméneutique et la déconstruction," included in Thomas Bedorf und A. B. Blank (eds.), *Diesseits des Subjektprinzips: Körper – Sprache – Praxis, Sophist: Sozialphilosophische Studien*, Vol. 3 (Berlin: Edition Humboldt, 2002), pp. 211–27. A large part of Chapter 6 has been adapted from the article "Complementarity and Deconstruction: Plotnitsky's Analysis and Beyond," *Configurations*, **12**, No. 3 (2007), 435–76. The main arguments in Chapters 4 and 6 are also presented in a more condensed form in my Japanese-language paper "Bohr no sōhosei: sono kōzō to tetsugakuteki gan'i [Bohr's Complementarity: Its Structure and Philosophical Implications]," *Shisō*, No. 986 (2006), 27–47.

This book could not have come into being without help, support, and encouragement from many people. First of all, I express my profound gratitude to my Ph.D. supervisor Hans Radder and co-supervisor Angela Roothaan of the Vrije Universiteit for their excellent and very thoughtful guidance of my project, which forms the basis of the present work. Hans Radder has, moreover, been so generous as to continue – far beyond the period of the doctoral project – giving me valuable advice for further advancement of my research. I am also particularly indebted to Arkady Plotnitsky for stimulating discussions on complementarity and deconstruction, especially during the years 2001–2002, in which period he kindly accepted me as a visiting scholar at Purdue University, USA. My sincere thanks are also due to Gregor Schiemann for fruitful dialogues and exchanges on quantum theory and the Bohr–Heisenberg relation on various occasions, not least during my stay at his institute at Wuppertal University, Germany, in 2009. I owe a special debt of gratitude to Jan Faye, Tetsuya Takahashi, and Osamu Kanamori, who kindly answered my questions and gave me further suggestions regarding Bohr's thought, Derridean deconstruction, and the debate known as the Science Wars, respectively. I am also grateful to the anonymous reviewer of Springer for her/his careful and sympathetic report, which greatly helped me complete the final manuscript. Further thanks should go to many colleagues and friends, including Kaoru Aoki, Karen Barad, Andrew Feenberg, Shingo Fujita, Kenji Ito, Peter Kirschenmann, J. Murray Murdoch, Jr., Don Nilson, Henk de Regt, László Ropolyi, Joseph Rouse, Erhard Scholz, Patrocinio Schweickart, Gábor Szabó, and Kiichi Tachibana, for their helpful comments, suggestions, and constructive criticisms at various stages of my project. I am no less thankful to the people who carefully checked and helped improve the English, especially Alan Farr, Don Nilson, Lawrence R. Pfleger, R. Jeffrey Ringer, and Michael A. Santone, Jr. I owe gratitude to the editors at Springer, above all Charles Erkelens and Lucy Fleet, for their interest in my book proposal and their sustained effort and friendly cooperation in the whole process leading up to the publication.

Finally, my special thanks go to Atsuko Sugawara for her immeasurable support and encouragement in daily life as well as to my mother Megumi and my late father Hiroshi for their constant warm support over the years. Looking back on the past, it was my father, an elementary particle physicist, who was the first in my childhood to arouse my interest in natural science and later, in my student days, to draw my attention to philosophical questions of quantum theory. Unfortunately, however, he did not live to see the completion of the work that owes him so much. This book is dedicated with deep gratitude to the memory of him, Hiroshi Katsumori.

Contents

Introduction

In the second half of his academic career, the Danish physicist Niels Bohr (1885–1962) developed philosophical thought revolving around the concept of "complementarity." Bohr's idea of complementarity is primarily concerned with quantum theory, his major field of research, but extended to other fields of knowledge and experience as well. Since the rise of quantum theory itself was closely intertwined with philosophical questions concerning such basic notions as subject and object, space-time and causality, chance and necessity, and so on, it is no wonder that Bohr's complementarity has philosophical implications reaching far beyond the field of atomic physics.[1] The question may be posed, however, whether these implications have so far been adequately understood.

Bohr often characterized his idea of complementarity by the metaphorical dictum that 'we are both onlookers (or spectators) and actors in the great drama of existence.' In his view, while modern physical science has hitherto sought to see nature from the standpoint of a pure 'spectator,' the development of quantum theory has suggested that there can be no such purely detached standpoint, and that scientists themselves, as it were, unavoidably get involved in the drama of nature. More specifically, the observation of an atomic object carries with it an unavoidable and uncontrollable interaction with the measuring instrument – a circumstance which puts in question the conventional notion of independent objective reality. This led Bohr to develop the idea that space-time coordination and the claim of causality (or the use of the momentum-energy conservation laws) stand – as an 'actor' and a 'spectator,' respectively – in a "complementary," that is, mutually exclusive and yet jointly completing relationship. Further, he extended this idea of complementarity to fields outside quantum theory, pointing to the complementary relation between access to the typical aspects of life and its physical analysis, between psychical experience and its reflective analysis, and so forth. In this way, he presented the idea of complementarity as a basic conceptual framework not only for quantum theory, but covering wide-ranging fields of knowledge and experience.

[1] Throughout this study, I follow Bohr himself in using the term 'complementarity' to refer not only to his *concept* of complementarity, but also, in certain contexts such as the present one, to his overall philosophical thought revolving around this concept.

On closer inspection, however, Bohr's complementarity will prove to be neither structurally homogeneous nor historically unchanged. It appears to me that, in the 'early' period starting in 1927,[2] his complementarity argument with regard to quantum theory centers on the notion, which I call static, that we are 'spectators' and 'actors' only in different and separate situations. On the other hand, his account of complementarity in psychology and epistemology during the same period contains the dynamic notion that we are 'actors' precisely *in being 'spectators.'* It will further turn out that subsequently, in and through his 1935 debate with Einstein and his collaborators, Bohr carried over this dynamic conception of complementarity into his very interpretation of quantum theory, thus tacitly but significantly reorganizing his conceptual framework. That is, the use of the momentum-energy conservation laws, initially associated only with a 'spectator's' detachment, was reinterpreted as having the character of an 'actor's' involvement as well, so that the dynamic relation between the roles of 'spectators' and 'actors' became all-embracing.

Regrettably, in my view, neither the structural complexity nor the historical development of Bohr's complementarity as just briefly suggested has been fully or appropriately treated by many commentators. As regards the structural complexity, von Weizsäcker's early (in part valid) distinction between "parallel" and "circular" complementarity (1976, 284, 294), in particular, has not been taken seriously enough or further elaborated by others. In the historical dimension, although recent studies on Bohr's thought have increasingly addressed its development over time, their arguments are limited by their mostly unquestioned premise that his philosophical position and its possible diachronic changes are to be found along the conventional axis of realism and anti-realism. Generally speaking, many prior studies have sought to situate Bohr's complementarity within contexts formed by the analytic or 'mainstream' philosophy of science. In these contexts, his thought has often been interpreted under tacit presuppositions – specifically the "unambiguous meaning" of concepts or words, and definite philosophical positions such as realism and anti-realism – which are precisely of the kind targeted by his epistemological critique. This problem seems also to be connected with the widespread underestimation, or even the sheer disregard, of his complementarity argument outside quantum physics – a part of his work which would strongly suggest the need to broaden or transform the common interpretive schemes. In this way, many commentators' insufficient recognition of the different, and in part conflicting, layers of Bohr's thought as well as the different phases of its development appears to be correlative with what I see as the narrow or inadequate conceptual frames of their overall interpretive approaches.

Designed to exceed these limitations, the present work is characterized by the following set of interrelated methodological features. First, as suggested above, I

[2] By the 'early' period I mean not the early phase of Bohr's whole career, but that of the development of his idea of complementarity starting in 1927. The designations 'middle' and 'late' are also used in a parallel way.

focus on the complex conceptual structure of Bohr's complementarity as comprising different modes of the 'spectator–actor' relationship. Specifically, I not only distinguish between the static and the dynamic conceptions of complementarity, but further subdivide the former into what I call static-contrastive and static-symmetrical conceptions. By doing so, I seek to elucidate the heterogeneous roles played by these distinct layers of his thought in questioning and transforming traditional scientific-philosophical notions. Second, this analysis is closely coupled with an investigation into the historical process through which Bohr's complementarity underwent significant conceptual changes. Dividing the whole period in question into three consecutive phases, I will argue, in particular, that the transition from the 'early' to the 'middle' period is characterized by an extension of his dynamic conception of complementarity from non-physical fields to quantum theory, while this dynamic conception later paradoxically helped produce the static-symmetrical conception. It will further be shown that, in the 'late' period, this static-symmetrical conception became predominant, serving as a basis for his eventual attempt to restore the standpoint of a pure 'spectator.' The third aspect of my study concerns situating Bohr's thought in a broader recent and contemporary philosophical context. In particular, I explore the possible conceptual links between Bohr's complementarity and hermeneutic philosophy, specifically between the 'spectator–actor' relation in the former and what Gadamer and Ricoeur characterize as the relation between "belonging" and "alienation" or "distanciation." Further, I examine the conceptual intersections of complementarity with Derridean deconstruction as well, proceeding through a critical appraisal of Plotnitsky's prior analysis of the subject. This series of inquiries will hopefully not only contribute to a better understanding of Bohr's complementarity, but also help bridge the gulf between the 'main-stream' philosophy of science and important branches of contemporary philosophy hitherto little associated with physical science. In other words, this work is designed not simply as a new approach to Bohr's thought, but, by extension, as a step toward overcoming the still persisting conceptual barriers between different orientations of contemporary philosophy.

The present work consists of six chapters, preceded by this Introduction and followed by Concluding Remarks. After briefly tracing in Chapter 1 the historical rise of quantum theory and Bohr's contribution to it, I proceed in Chapter 2 to survey the development of his idea of complementarity during the above three periods from 1927 until the end of his life. In Chapter 3, I review a number of prior interpretations of complementarity – some of which have important bearing on later discussions – by physicists as well as historical and philosophical commentators. Starting from and with reference to the findings in these earlier chapters, Chapter 4, arguably the most pivotal of the whole work, is devoted to a philosophical-historical analysis of Bohr's complementarity from my own interpretive point of view as sketched above. Based on the outcome of this analysis, in Chapters 5 and 6 I further discuss the relation of complementarity to hermeneutic philosophy and Derridean deconstruction, respectively, pointing out both conceptual similarities and divergences. Readers already familiar with, or not particularly interested in, the historical development of quantum theory can skip Chapter 1 without having essential difficulty in understanding the philosophical import of the following chapters. Chapter 3 may also be skipped in the first reading, although my accounts of prior studies should serve to elucidate my own interpretive standpoint on Bohr's thought. Further, readers who are already well versed in Bohr's texts and want to enter quickly into the core arguments of this work could even start with Chapter 4 and, when occasion demands, move back to earlier chapters.

Chapter 1
Bohr and the Development of Quantum Theory: A Brief Review

Notwithstanding the broad philosophical implications of Niels Bohr's thought reaching far beyond the confines of physics, quantum theory was from the beginning, and remains to be, the field of primary importance with which his idea of complementarity is concerned. Understanding of his thought therefore requires at least a brief survey of quantum physics as it historically developed in the early twentieth century. In this opening chapter, I wish accordingly to offer a sketch of the development of quantum theory – including the central role played by Bohr – within which his path toward the idea of complementarity is situated.

The rise of quantum theory is commonly traced back to Max Planck's work on radiation at the turn of the century. In 1900, in an effort to account for apparent anomalies in black-body radiation, Planck proposed a "purely formal assumption": The energy radiated by the black body should be treated as composed of small but finite "elements of energy" (1901, 556), which would later be called "energy quanta."[1] In quantitative terms, he suggested that each of these 'elements of energy,' E, is proportional to the frequency of vibration ν of the resonators: $E = h\nu$, where h is a constant (1901, 561), which would later be known as "the quantum of action" or simply as Planck's constant. Although Planck himself meant this hypothesis as no more than a provisional device and wished to preserve the continuity of radiation and of the motion of the oscillating resonators producing radiation,[2] his quantum hypothesis in effect initiated a radical innovation of modern physical science (see PWNB, 1:28).

In 1905, Albert Einstein took the next important step in developing the idea of quanta. Guided by an analogy between black-body radiation and certain properties of an ideal gas as described by statistical mechanics, Einstein proposed the following "heuristic viewpoint": Radiation may be regarded as small packets of energy, which he called "light quanta" (later "photons"), with the amount of energy being

---

[1] According to Thomas S. Kuhn, Planck's change of vocabulary from "element" to "quantum" signals a change in meaning of the quantity $h\nu$ "from a mental subdivision of the energy continuum to a physically separable atom of energy" (Kuhn 1978, 363, cf. 201).

[2] Against the conventional understanding, Kuhn argues that not in 1900, but only in 1908, did Planck realize that his hypothesis "demanded discontinuity" (Kuhn 1978, 355, cf. viii, 125–30, 196–202). See also Jammer (1989, 35f.) and Darrigol (1992, xviii, 70ff.).

M. Katsumori, *Niels Bohr's Complementarity*, Boston Studies in the Philosophy of Science 286, DOI 10.1007/978-94-007-1748-0_1,

proportional to the frequency of the radiation (1905, 144; see PWNB, 1:29).[3] While Planck's hypothesis of discrete energy was restricted to the resonators producing radiation, Einstein proposed the quantization of radiation itself. Einstein did not, however, intend this "light-quantum hypothesis" simply to replace the wave theory of light. Rather, in his view, one needed a new theory uniting both the continuous and the discontinuous aspects of radiation or, in other words, accounting for the wave-particle duality of light. Although this idea of wave-particle duality, associated with the light-quantum hypothesis, was not commonly accepted until around the mid-1920s (see Kuhn 1978, 180, 187), it anticipated a significant phase of the later development of quantum theory to be seen below. Planck's and Einstein's contributions taken together, we can say that the idea of the quantum thus introduced to physics not only constituted a challenge to the wave theory of light,[4] but rather that its innovative role lay in the introduction of a kind of discontinuity that was foreign to the classical theories. Classical physics, constituted above all by Newton's laws of motion and Maxwell's equations of electromagnetism, basically assumed the continuity of motions, fields, and waves. In contrast, quantum theory, which may be (retrospectively) viewed as initiated by the above two physicists, brought in discontinuity as an essential feature of physical processes.

A further development of quantum theory proceeded in close connection with inquiries into the atomic structure of matter. In 1911, based on experimental results, Rutherford presented the novel idea that most of the atomic mass is localized in a small, positively charged nucleus, around which negatively charged electrons move (see PWNB, 1:30, 3:30ff.). This atomic model of Rutherford served as the starting point from which Niels Bohr developed his own seminal atomic theory with the aid of the idea of quanta.[5] Recognizing a serious theoretical problem facing Rutherford's model – the problem of accounting for atomic stability in classical terms – Bohr took a step beyond the limits of the classical theories by recourse to Planck's hypothesis. In a 1913 trilogy of papers entitled "On the Constitution of

[3] Einstein suggested that the 'photoelectric effect' and some other light phenomena may be accounted for by the light-quantum hypothesis, although the photoelectric effect did not constitute his principal concern in this work (see 1905, 144ff.). See also Jammer (1989, 22), Kuhn (1978, 221), and Murdoch (1987, 5f.).

[4] Although since early modern times the nature of light was a controversial issue, it was in the nineteenth century that the wave theory of light became predominant by virtue of the discovery that light exhibits the phenomena of diffraction and interference. This wave theory of light was further developed by James Clerk Maxwell, whose theory of electromagnetism offered an elaborate account of light as an electromagnetic wave (see PWNB, 1:27). With the general acceptance of Maxwell's theory, the wave nature of light appeared to be firmly established before the advent of quantum theory.

[5] Bohr stayed in Britain for postdoctoral research from 1911 to 1912. After working at J. J. Thomson's laboratory in Cambridge, he transferred in early 1912 to Manchester University to work with Rutherford. See Pais (1991, 117–31).

Atoms and Molecules,"[6] he began by introducing the following assumptions: First, electrons inside an atom are located in specific discrete orbits, each corresponding to a well-defined energy, and the dynamical equilibrium of the systems in these "stationary states" is "governed by the ordinary laws of mechanics" (NBCW, 2:232). Second, however, instead of continuously emitting radiation during their motion in a stationary state, electrons emit or absorb radiation only when they move from one stationary state to another. Radiation thus emitted or absorbed is "homogeneous," and its frequency ν is, in accordance with the quantum hypothesis, given by the relation $E_1 - E_2 = h\nu$, where E_1 and E_2 are the energies of the two states in question (NBCW, 2:167f.; see PWNB, 1:31). On these two assumptions, combined with an additional quantum rule,[7] Bohr calculated the frequencies of radiation emitted by the hydrogen atom, in particular, and showed the result to be in a remarkable agreement with what was empirically known as the Balmer formula for the line spectrum (see PWNB, 1:32).

As suggested above, in the 1913 trilogy, Bohr distinguished between two areas, the one in which classical physics remains valid and the one in which it no longer does. Classical physics applies only to the dynamical equilibrium of the systems in stationary states, and not to "the passing of the systems between different stationary states" (NBCW, 2:167). This implies, in particular, that the mechanical and electrodynamical stability of the atomic model is incomprehensible from a classical point of view.[8] At the same time, however, Bohr introduced a crucial link between the classical and the quantum areas. In one of the formulations of the theory, he assumed that, in the limiting region of high quantum numbers, the results of classical electrodynamics and his quantum theory asymptotically coincide. Specifically, "the frequency of the radiation emitted during the passing of the system between successive stationary states will coincide with the frequency of revolution of the electron in the region of slow vibrations" (NBCW, 2:174). This agreement (for large quantum numbers) between the quantum frequency of the light emitted and the classical frequency associated with the motion of the electron constitutes the initial form of what would be known as the "correspondence principle."

[6] Niels Bohr, "On the Constitution of Atoms and Molecules," *The Philosophical Magazine* 26 (1913): 1–25, 476–502, 853–75; reprinted in NBCW, 2:161–233.

[7] As Max Jammer comments, in the trilogy Bohr gives three different formulations of the quantum rule (Jammer 1989, 78ff.), the simplest of which assumes that "the ratio between the total energy, emitted during the formation of the configuration, and the frequency of revolution of the electron is an entire multiple of $h/2$." In the case of a circular orbit, this is equivalent to the assumption that "the angular momentum of the electron round the nucleus is equal to an entire multiple of $h/2\pi$" (NBCW, 2:233; cf. 164f., 184f.). See also Darrigol (1992, 86ff.).

[8] Furthermore, the radiation mechanism in Bohr's theory differs essentially from that in classical physics: According to classical electrodynamics, the basic frequency and the higher harmonics are emitted in one atomic process, while in Bohr's theory every quantum frequency originates from a separate atomic transition. See Radder (1988, 130f.).

From then on until the mid-1920s, Bohr sought further to develop quantum theory with the aid of this idea of "agreement" or "analogy"[9] between quantum and classical theories.[10] Specifically, he succeeded in accounting for the intensities and polarizations – in addition to the frequencies – of spectral lines by extending the idea of 'agreement' to smaller quantum numbers as well.[11] In 1920, he introduced the term "correspondence" as follows:

> [...] although the process of radiation can not be described on the basis of the ordinary theory of electrodynamics [...] there is found, nevertheless, to exist a far-reaching *correspondence* between the various types of possible transitions between the stationary states on the one hand and the various harmonic components of the motion on the other hand (NBCW, 3:245f.; see PWNB, 1:37).

This correspondence is such that "the present theory of spectra is in a certain sense to be regarded as a rational generalization of the ordinary theory of radiation" (NBCW, 3:246).[12] Bohr thus formulated the correspondence principle as a general principle of analogy between quantum and classical theories, and, more specifically, between the properties of the radiation emitted by electrons and the Fourier components of the motion of these electrons (see Radder 1988, 134). The development of quantum theory with the aid of this correspondence principle did not, however, proceed without difficulty. Rather, for the time being, the theory failed to achieve coherence, and was faced with serious problems such as that of determining the energy states of the helium atom, the anomalous Zeeman effect, and so forth.

As mentioned earlier, Einstein's idea of light quanta was not accepted promptly by the community of physicists. This situation began to change in 1922, however, when A. H. Compton presented experimental results which strongly suggested the particle character of light. In his experiments on atoms irradiated with X-rays, he observed the way in which the radiation was scattered with changes in wavelength and direction. He was able to account for these changes on the assumption that light quanta of the incident radiation collide with electrons in the scattering substance. This explanation of the Compton effect in terms of light quanta constituted a decisive factor leading to a general recognition of the particle character of radiation. Yet this did not take place immediately, and Bohr, in particular, still refused to accept the

[9] In 1913, Bohr spoke of "the most beautiful analogi [*sic*] between the old electrodynamics and the considerations used in my paper." Letter from Bohr to Rutherford, March 21, 1913 (NBCW, 2:584f.). Further, in 1917, he used the expression: "the formal analogy between the ordinary theory of radiation and the [quantum] theory" (NBCW, 3:100). See Chevalley (1995, 15).

[10] Bohr began to extend his idea of 'agreement' by combining it with generalized quantum conditions introduced by Sommerfeld in 1915 (see PWNB, 1:38).

[11] In Hans Radder's account, the development of Bohr's correspondence principle may be divided into three consecutive phases: the first (1913–1915), the second (1916–1922), and the third (1923–1925). While the initial form of correspondence as Bohr conceived it in the first phase concerned numerical agreement or "agreement of calculations," he came in the second phase to assume also conceptual continuity between his quantum theory and the classical theories (Radder 1988, 129, 133; cf. 1996, 55).

[12] In 1923, Bohr characterized the correspondence principle as a "general law holding for all quantum numbers" (NBCW, 3:577).

idea of light quanta. In 1923, Bohr remarked that the hypothesis of light quanta can in no way be regarded as a "satisfactory solution," because it "introduces insuperable difficulties, when applied to the explanation of the phenomena of interference," and that its underlying picture "excludes in principle the possibility of a rational definition of the conception of a frequency ν" (NBCW, 3:492; see Honner 1987, 37ff.).

Faced with general difficulties challenging quantum theory, Bohr made a move away from intuitive space-time pictures by replacing his original orbital model of the atom with a virtual field model (see NBCW, 5:115). In a 1924 joint paper on the interaction between radiation and matter,[13] Bohr, Kramers and Slater assumed the atom to be a set of "virtual harmonic oscillators." In their view, every atom in a stationary state "communicate[s] continually with other atoms" by means of "virtual radiation fields" produced by the oscillators, whose frequencies correspond to "the various possible transitions to other stationary states" (NBCW, 5:106).[14] In order to reconcile the notion of continuous electromagnetic fields with that of discontinuous quantum transitions, however, the authors held that the energy and momentum exchanged between atoms and radiation are not strictly conserved in individual emission and absorption processes, but only on the average over numerous transitions. That is, Bohr and his collaborators abandoned "a direct application of the principles of conservation of energy and momentum," reducing these principles to statistical laws (NBCW, 5:107; see Jammer 1989, 188f.).[15]

Shortly afterward in the same year, however, Bothe and Geiger presented experimental results indicating that, in the interaction between radiation and matter, the conservation laws strictly hold. This forced Bohr to abandon the above Bohr-Kramers-Slater theory as such,[16] but still did not lead him to accept literally the hypothesis of light quanta. What the Bothe-Geiger results suggested for Bohr is not

---

[13] Niels Bohr, Hendrik A. Kramers, and John C. Slater, "The Quantum Theory of Radiation," *The Philosophical Magazine* 47 (1924): 785–802; reprinted in NBCW, 5:101–18 as well as in van der Waerden (1967, 159–76). Kramers was Bohr's assistant at his Institute for Theoretical Physics, while Slater made a short visit at the institute in early 1924 (see Murdoch 1987, 24f.). After being founded in 1921, the Institute for Theoretical Physics of Copenhagen University served as an international center of the research of quantum theory in which many foreign physicists stayed and collaborated with Bohr and with each other. See Pais (1991, 171).

[14] According to Radder, Bohr's correspondence principle at this third and final stage assumes an agreement only for large quantum numbers, and a formal and numerical correspondence rather than a conceptual continuity between classical and quantum theories (Radder 1988, 140, 143). That is, this principle came again to concern a non-conceptual correspondence, but, unlike in the first stage, took on a formal character in distancing itself from intuitive space-time pictures. See also Radder (1996, 55f.).

[15] In this Bohr-Kramers-Slater paper, the authors still assumed a critical attitude toward the light quantum hypothesis by saying that, despite its "great heuristic value," "the theory of light-quanta can obviously not be considered as a satisfactory solution of the problem of light propagation" (NBCW, 5:103).

[16] Bohr remarked: "Since now a unique coupling of atomic processes seems actually to be a fact even for radiation phenomena, the approach taken in this paper [based on the Bohr-Kramers-Slater theory] must probably be abandoned" (NBCW, 5:205).

the discrete image of light quanta, but rather a breakdown of the classical space-time description of the interaction between radiation and matter (see Chevalley 1994, 37). In 1925, Bohr remarked that "the assumption of a coupling between changes of state in distant atoms by radiation excludes the possibility of a simple description of the physical events in terms of intuitive (*anschaulich*) pictures," so that "we must take recourse to symbolic analogies to a still higher degree than before."[17] Later in the same year, he indeed spoke of "an essential failure of the pictures in space and time on which the description of natural phenomena has hitherto been based" (PWNB, 1:34f.; cf. 50; NBCW, 5:190/204, 193/206).[18]

Around the mid-twenties, the development of quantum theory came to a crucial stage, notably by virtue of the young Werner Heisenberg's contribution. In 1925, having just started research under Bohr's guidance,[19] Heisenberg investigated how the quantum numbers and energy states in an atom are connected with the experimentally obtained frequencies and intensities of the light spectra.[20] In so doing, he chose to "discard all hope of observing hitherto unobservable quantities (such as the position and period of the electron)," and instead sought to "establish a theoretical quantum mechanics, analogous to classical mechanics, but in which only relations between observable quantities occur" (1984ff., A1:383; cf. A1:503).[21] Starting from Bohr's correspondence principle (see PWNB, 1:49), Heisenberg reinterpreted the relations between the Fourier amplitudes of the position coordinate of the oscillator as quantum relationships between the observable properties of the emitted radiation. Later in the same year, in collaboration with Born and Jordan,[22] Heisenberg further elaborated his theory as a system of matrix calculus in which the dynamical variables were represented by infinite matrices instead of ordinary numbers. This theory, which would be named matrix mechanics, is thus marked by a discrete descriptive approach doing without any visualizable model for atomic processes (see Beller 1999, 19f.).

Concurrently with the rise of matrix mechanics, another, apparently independent, conceptual development took place, which goes back to Louis de Broglie's idea of matter waves (see Jammer 1989, 242). In 1923, inspired by Einstein's idea of the

[17] Letter from Bohr to Born, May 1, 1925 (NBCW, 5:310f./85, on 311/85; cf. 5:113; PWNB, 1:36). See also Petruccioli (1993, 127) and Chevalley (1995, 15f.).

[18] See Jammer (1974, 91). Bohr further maintained that a generalization of classical electrodynamics would "require a fundamental revolution in the concepts upon which the description of nature has been based until now" (NBCW, 5:191/205).

[19] Heisenberg stayed as a research fellow at Bohr's Institute for Theoretical Physics from September 1924 to April 1925, and again – this time as university lector and assistant to Bohr – from May 1926 to June 1927. See Pais (1991, 263) and Cassidy (1992, 183ff.).

[20] Werner Heisenberg, "Über die quantentheoretische Umdeutung kinematischer und mechanischer Beziehungen," *Zeitschrift für Physik* 33 (1925): 879–93; reprinted in Heisenberg 1984ff., A1:382–96; Eng. trans. "Quantum-Theoretical Re-Interpretation of Kinematic and Mechanical Relations," in van der Waerden 1967, 261–76.

[21] Trans. van der Waerden (1967, 262).

[22] It was Max Born who realized that the multiplication rule for oscillator amplitudes found out by Heisenberg was nothing other than the basic rule for multiplying matrices. See Jammer (1989, 215).

wave-particle duality of light, de Broglie had developed the novel view that the propagation of a wave may be associated with the motion of any material particle. Going beyond the conventional corpuscular view of matter, he had thus extended wave-particle duality to electrons and to matter in general.[23] In 1926, starting from this concept of matter waves, Erwin Schrödinger presented a version of quantum theory which describes the state of the atomic system in terms of a "wave function" ψ, and, in contrast to Heisenberg's matrix mechanics, retains the idea of visualizing atomic phenomena. On the basis of the wave function, which is governed by a differential "wave equation," Schrödinger sought to restore a physics of continuous processes.[24] Specifically, he attempted to describe all particles as the superposition of waves, and rejected the concept of discontinuous 'quantum jumps' (see 1926, I:375). While proving the mathematical equivalence between his wave mechanics and Heisenberg's matrix mechanics, Schrödinger held his theory to be superior to the latter precisely because of its capability to give a continuous and intuitive picture of atomic phenomena (see Jammer 1989, 270f.; Chevalley 1995, 16f.).

Bohr was willing to accept this wave mechanics of Schrödinger, specifically his wave-packet representation of material particles, while not agreeing with his intuitive understanding of waves (see Murdoch 1987, 43). This move is closely connected with a change in Bohr's attitude toward the problem of wave-particle duality. In 1925, he had still resisted accepting the light quantum hypothesis and wave-particle duality by maintaining that "[t]he formal nature of this statement [i.e. Einstein's light quantum hypothesis] is evident because the definition and measurement of [the] frequency [of light] rests exclusively on the ideas of the wave theory" (PWNB, 1:29; cf. 33, 46). In the next year, however, Bohr remarked that "it is interesting to see how the concept of wave or particle presents itself as the more appropriate concept depending on the place in the description where the assumption of discontinuity explicitly appears."[25] Further, in early 1927, he noted more explicitly that "[t]he representation of an electron by a group of de Broglie waves is [. . .] closely analogous to the representation of a light quantum by a group of electromagnetic waves."[26] This suggests that he now considered light quanta to be particles as 'real' as electrons, and attributed as much reality to de Broglie matter waves as

[23] Bohr at the time considered de Broglie's hypothesis of matter waves to be purely formal and devoid of realistic significance, just as Einstein's theory of light quanta. It is not until 1927 that the idea of matter waves would be supported by Davisson and Germer's experiments, which offered striking evidence of electron diffraction. See Murdoch (1987, 35f., 53).

[24] In the same year, 1926, Max Born offered an interpretation of the wave function ψ different from Schrödinger's own. According to Born, the square of the wave function, $|\psi|^2$, represents the probability of the associated particle's presence. See Jammer (1989, 301ff.).

[25] Letter from Bohr to Schrödinger, December 2, 1926 (NBCW, 6:462f., on 462). See Murdoch (1987, 46). On the role of Bohr's dialogue with Schrödinger in the formation of his idea of complementarity, see Catherine Chevalley's "Introduction" to Bohr (1991, 17–147, on 66ff.). See also Beller (1999, 122ff.).

[26] Letter from Bohr to Einstein, April 13, 1927 (NBCW, 6:421/23).

to electromagnetic waves. This is not to say, however, that he accepted literally these intuitive pictures of particles and waves for either radiation or matter. Rather, he reconceived electrons and light quanta – or electromagnetic waves and matter waves – equally as idealizations to which nothing in our perceptual experience exactly corresponds (see Murdoch 1987, 53).

It is along this line of thought and thus in a certain limited sense that Bohr accepted Schrödinger's continuity-oriented wave mechanics. In more general terms, Bohr maintained that we are left "only with the choice between Charybdis and Scylla, according to whether we direct our attention towards the continuous or the discontinuous aspect of the description."[27] The notion of continuity, which underlies the concept of causality, is indeed indispensable for "the definition of every concept" (NBCW, 6:462), and yet does not have unlimited validity.[28] On the other hand, despite his high regard for Heisenberg's achievement in matrix mechanics (see PWNB, 1:70f.), Bohr did not exactly share the young physicist's interpretation of his own theory, especially his ontological premise of discontinuity. In 1927, Bohr said that, rather than "only talk about particles and quantum jumps," we should "keep in mind how indispensable are the concepts of the continuous field theory in the present stage of science."[29] In this way, he held that neither wave mechanics nor matrix mechanics can be accepted as intuitive pictures of nature. Rather, both forms of quantum mechanics should be viewed as correspondence theories, that is, theories based on formal or symbolic analogies (see PWNB, 1:44, 75, 110f.).[30] As he put it: "on the basis of wave mechanics one can build a correspondence theory, which is in itself just as closed as matrix mechanics, which in its turn may be conceived as a correspondence theory resting on particle mechanics."[31] As we have seen earlier, around 1925, Bohr had come to the conviction that it is impossible to describe physical events simply in terms of intuitive space-time pictures.[32] Coupled with this, his new approach to the problem of continuity and discontinuity as outlined above formed the conceptual starting point from which he would soon develop the idea of complementarity (see Jammer 1974, 91).

In 1927, Heisenberg derived from matrix mechanics a set of relations characterized as "uncertainty" or "indeterminacy (*Unbestimmtheit*)," which would prove

[27] Letter from Bohr to Einstein, April 13, 1927 (NBCW, 6:418–21/21–24, on 419/21). Just before the quotation in the same letter, Bohr also wrote: "how intimately the difficulties of quantum theory are connected with the concepts or rather the words that are used in the customary description of nature, and which all have their origin in the classical theories."

[28] In Bohr's account, "the definition of every concept or rather every word presupposes the continuity of the phenomena and hence becomes ambiguous as soon as this presupposition cannot be upheld." Letter from Bohr to Schrödinger, December 2, 1926 (NBCW, 6:462).

[29] Letter from Bohr to Einstein, April 13, 1927 (NBCW, 6:421/23).

[30] See Murdoch (1987, 44), Chevalley (1995, 17), and Beller (1999, 35).

[31] Letter from Bohr to Schrödinger, December 2, 1926 (NBCW, 6:462). See Held (1994, 890).

[32] In 1925, commenting on the rise of matrix mechanics, Bohr said that "[i]n contrast to ordinary mechanics, the new quantum mechanics does not deal with a space-time description of the motion of atomic particles" (PWNB, 1:48; see Jammer 1974, 91).

to be of great significance for the interpretation of quantum theory.[33] He sought, in particular, to elucidate the meaning of the uncertainty relations with the aid of a thought experiment in which one observes an electron by illuminating it with light or gamma rays and receiving the scattered radiation with a microscope.[34] As an analysis of this experiment shows, "the more precisely the position is determined, the less precisely the momentum is known, and conversely."[35] This relation may be quantitatively expressed as $\Delta x \cdot \Delta p \sim h$, where Δx and Δp are the uncertainties of position x and momentum p, respectively, and h Planck's constant (Heisenberg 1984ff., A1:481).[36] Heisenberg also shows that a parallel relation holds for time t and energy E: $\Delta t \cdot \Delta E \sim h$ (1984ff., A1:485).[37]

While recognizing the crucial importance of this uncertainty work by Heisenberg, Bohr again did not entirely agree with his physical and philosophical views on the subject. In deriving the uncertainty relations, Heisenberg took a purely corpuscular approach, seeking to dispense with the wave concept.[38] Bohr countered this orientation by arguing that the wave-particle duality should be among the basic assumptions of quantum theory.[39] As regards Heisenberg's thought experiment mentioned above, Bohr maintained that what prevents measurement of the momentum change is not simply its discontinuity, but the indispensability of the wave model for interpreting the experiment (see Murdoch 1987, 49). In his view, the uncertainty relations are a special case of the state of affairs that he would soon formulate as complementarity.[40]

---

[33] Werner Heisenberg, "Über den anschaulichen Inhalt der quantentheoretischen Kinematik und Mechanik," *Zeitschrift für Physik* 43 (1927): 172–98; reprinted in Heisenberg (1984ff., A1:478–504); Eng. trans. J. A. Wheeler and W. H. Zurek, "The Physical Content of Quantum Kinematics and Mechanics," in Wheeler and Zurek (1983, 62–84).

[34] According to some commentators, as is suggested by the title of the paper, Heisenberg here restored the notion of visualizablity or intuitiveness (*Anschaulichkeit*) – which he had once abandoned in 1925 – through a "subtle redefinition" of the term *anschaulich* (Camilleri 2009, 49; see Beller 1999, 49, 67ff.).

[35] In Heisenberg's account, the change in the momentum of the electron "is the greater the smaller the wavelength of the light employed – that is, the more exact the determination of the position." Since this change itself cannot be determined, "[a]t the instant at which the position of the electron is known, its momentum [. . .] can be known up to magnitudes which correspond to that discontinuous change" (1984ff., A1:481; trans. Wheeler and Zurek 1983, 64).

[36] Trans. Wheeler and Zurek (1983, 64).

[37] Trans. Wheeler and Zurek (1983, 68).

[38] As Heisenberg recalled later, at the time he still considered Schrödinger's wave mechanics "as an extremely useful tool for solving the mathematical problems of quantum mechanics, but not more" (1967, 102).

[39] In a letter from around the same period, Bohr noted that "the uncertainty mentioned is not only connected to the presence of discontinuities, but also to the very impossibility of a detailed description in accordance with those properties of material particles and light that find expression in the wave theory." Letter from Bohr to Einstein, April 13, 1927 (NBCW, 6:418–21/21–23, on 421/23).

[40] See Heisenberg (1967, 106). Similar to his 1925 work on quantum mechanics, Heisenberg's 1927 uncertainty paper also tends toward a positivist view of science according to which "physics

In this first chapter, we have briefly traced the development of quantum theory in the early twentieth century up until 1927 – the year of Heisenberg's formulation of the uncertainty relations *and* Bohr's subsequent introduction of complementarity. It is not simply the case that Heisenberg's uncertainty work gave rise to Bohr's idea of complementarity. Rather, the two physicists' approaches to quantum theory were so closely interrelated that their intense discussion and debate led Bohr to elaborate further his own views on the subject, which took shape as the argument of complementarity. In the next chapter, I will accordingly begin to address Bohr's complementarity, the overall theme of the present study, tracing its development from the year 1927 onward.

---

should only formally describe the relation between perceptions" (1984ff., A1:503; cf. 480, 491). As we will see later on, this positivist idea was not shared by Bohr, although their positions were often confounded by others. It is also noteworthy that, as suggested by some commentors, the early Heisenberg's positivist tendency may not so much constitute a definite philosophical commitment, but rather serve as a "post facto justification" of his approach to quantum mechanics (Camilleri 2009, 17; see also Beller 1999, 52ff.; Schiemann 2008, 45).

Chapter 2
An Overview of Bohr's Complementarity

As sketched in the previous chapter, Niels Bohr's engagement in the development of quantum theory, especially in its crucial phase around the mid-1920s, proceeded with a critical examination of basic concepts such as continuity/discontinuity, space-time and causality, which underlie not only atomic physics, but the modern-scientific view of nature in general. This led to his introduction in 1927 of the idea of complementarity as an interpretation of quantum theory, newly established as quantum mechanics. While closely linked with Heisenberg's uncertainty relations, Bohr's complementarity is marked by its own distinctive philosophical notions and implications. Here and throughout this study (as noted in the Introduction), I use the term 'complementarity' to refer not only to Bohr's *concept* of complementarity, but – following his own and many commentators' usage – also to his overall philosophical thought revolving around this concept.

In the present chapter, I will survey in outline the development of Bohr's idea of complementarity from 1927 until the end of his life. In so doing, as noted earlier, I divide the whole period into three consecutive phases: (1) the 'early' period, stretching from his 1927 Como lecture until – but not including – his 1935 debate with Einstein and his collaborators, (2) the 'middle' period, from this debate until around 1950, and (3) the 'late' period, from then on until his death in 1962. I employ this periodization tentatively here, while its justification will be given only by a detailed philosophical-historical analysis to be offered in Chapter 4. Sections 2.1 and 2.2 of this chapter are devoted to an overview of Bohr's 'early' complementarity with regard, respectively, to quantum theory and to other fields of experience such as biology and psychology. In Section 2.3, I proceed to the 'middle' period, most pivotal to the whole historical development of complementarity, and then finally in Section 2.4 to the 'late' period.

2.1 The 'Early' Period: Complementarity in Quantum Theory

It is in his lecture delivered in Como, Italy, in September 1927, entitled "The Quantum Postulate and the Recent Development of Atomic Theory," that Bohr for

M. Katsumori, *Niels Bohr's Complementarity*, Boston Studies in the Philosophy of Science 286, DOI 10.1007/978-94-007-1748-0_2,

the first time presented in public the term and idea of complementarity.[1] In this section, I outline Bohr's complementarity with regard to quantum theory, as it was developed during the 'early' period starting with the Como lecture. In so doing, while primarily drawing on his Como paper,[2] specifically its six-paragraph first section,[3] I seek to reconstruct his argument with reference also to other relevant works from the period such as "The Quantum of Action and the Description of Nature" and "Introductory Survey," both written in 1929 (PWNB, 1:92–101, 1–24, respectively).[4]

In his Como paper, Bohr starts his argument by noting that "our usual description of physical phenomena is based entirely on the idea that the phenomena concerned may be observed without disturbing them appreciably." This assumption – the accessibility of phenomena undisturbed by observation – underlies classical mechanics and is still maintained in the theory of relativity.[5] However, in Bohr's account, a new situation has arisen with the development of quantum theory, specifically with "the so-called quantum postulate, which attributes to any atomic process an essential discontinuity, or rather individuality, completely foreign to the classical theories and symbolized by Planck's quantum of action" (PWNB, 1:53; cf. 28, 93). This "individuality" implies that "any observation of atomic phenomena will involve an interaction with the agency of observation not to be neglected," thus introducing an unavoidable interference with the course of phenomena (PWNB, 1:54; cf. 11). In more specific terms, any observation aimed at space-time coordination brings about

[1] More specifically, the lecture was delivered at the International Congress of Physics in commemoration of the centenary of the death of Volta, on September 16, 1927. See Murdoch (1987, 55).

[2] I draw on the version published in *Nature* (Suppl.) 121 (1928): 580–90, and subsequently reprinted in Bohr's 1934 book *Atomic Theory and the Description of Nature* – later Volume I of *Philosophical Writings of Niels Bohr* (PWNB, 1:52–91). This version represents a considerable expansion of the text printed earlier in the conference proceedings, but retains the most basic ideas of the latter. See NBCW, 6:44, 110–12 and Faye (1991, 60).

[3] As Henry J. Folse comments, "Bohr presents virtually his entire argument in the six paragraphs which comprise the first section of the [Como] paper" (1985, 108).

[4] The relevant works also include the 1929 article "The Atomic Theory and the Fundamental Principles underlying the Description of Nature" (PWNB, 102–19) and "Faraday Lecture: Chemistry and the Quantum Theory of Atomic Constitution," published in 1932 (NBCW, 6:371–408). Incidentally, my reconstruction of Bohr's complementarity argument is more or less in accord with the following five-step summary by Max Jammer: "1. Indivisibility of the quantum action (quantum postulate)./ 2. Discontinuity (or individuality) of elementary processes./ 3. Uncontrollability of the interaction between object and instrument./ 4. Impossibility of a (strict) spatiotemporal and, at the same time, causal description./ 5. Renunciation of the classical mode of description" (1974, 101). Jammer's account seems to me, however, not to focus enough on the complementary relation between observation and the definition of the state (see 1974, 91).

[5] In Bohr's account, "the theory of relativity reminds us of the subjective character of all physical phenomena, a character which depends essentially upon the state of motion of the observer" (PWNB, 1:116; see Faye 1991, 166). Yet, in this theory, "the conception of the objective reality of the phenomena open to observation is still rigidly maintained." Bohr emphasizes that this classical ideal "cannot be attained in the description of atomic phenomena" (PWNB, 1:97; cf. 2:25, 41).

an exchange of momentum and energy with the measuring rods and clocks (see PWNB, 1:98). What Bohr considers crucial, however, is the fact that this interaction between the object and the instrument of observation is not simply unavoidable, but "uncontrollable," which represents a "feature of irrationality" brought into the description of nature (PWNB, 1:91; cf. 10, 54).[6] In classical physics, the observational interaction, even if not negligible, was regarded as in principle controllable, that is, capable of being theoretically determined so that its effect may be compensated for (see Folse 1985, 92). In quantum theory, by contrast, "just the impossibility of neglecting the interaction with the agency of measurement means that every observation introduces a new uncontrollable element" (PWNB, 1:68). For insofar as the measuring instrument is to fulfil its purpose, the interaction itself cannot be taken into account, and thus the magnitude of the disturbance always remains unknown (see PWNB, 1:11, 98).

This circumstance, in Bohr's view, has the following consequences. First, any space-time observation of the system brings with it "a complete rupture in the causal description of its dynamical behaviour" (PWNB, 1:68; cf. 98). That is, as he put it in 1929, "any observation takes place at the cost of the connection between the past and the future course of phenomena," thus depriving to a certain degree "the information given by a previous measurement of its significance for predicting the future course of the phenomena" (PWNB, 1:11, 18). Second, in Bohr's account, "an independent reality in the ordinary physical sense can neither be ascribed to the phenomena nor to the agencies of observation" (PWNB, 1:54). For, as he again put it explicitly in 1929, the observational interaction "prevents a sharp distinction being made between a phenomenon and the agency by which it is observed" (PWNB, 1:11).[7] In other words, we are faced in quantum theory with "the limit [. . .] of the possibility of speaking about phenomena as existing objectively" (PWNB, 1:115; cf. 93, 2:7). For this reason, he also maintains that "the concept of observation is in so far arbitrary as it depends upon which objects are included in the system to be observed" (PWNB, 1:54; cf. 67). We can thus see that, in Bohr's view, the uncontrollability of observational interaction entails a breakdown not only of causality, but also of the unambiguous notion of objective reality.

On the other hand, Bohr no less stresses the need to define the state of the atomic object, specifically its momentum or energy, and to describe it in terms of causality. This requires eliminating the observational influence on the phenomenon and maintaining "the distinction between object and agency of measurement" (PWNB, 1:68). In order to account for the stability of the atom, in particular, we must have "the conception of stationary states," which requires "a complete renunciation as regards a time description" and "the exclusion of all interactions with individuals not belonging to the system" (PWNB, 1:80). Taken together with the previous point, this leads to the paradoxical circumstance that the object cannot be separated from the agency

---

[6] Bohr at times qualifies this notion of irrationality by a phrase such as "irrational from the point of view of the classical theories" (PWNB, 1:7; cf. 19).

[7] For similar passages from the 'middle' period, see, for example, PWNB, 2:30 and 39.

of observation, while, on the other hand, it must be independent of the latter (see PWNB, 1:68). This state of affairs may be characterized more specifically as the conflict between space-time coordination by means of observation and the claim of causality associated with the definition of the state, or, more or less equivalently, between the use of space-time concepts and that of "the laws of conservation of energy and momentum" (PWNB, 1:11; cf. 61, 94).[8] As Bohr puts it in his Como lecture,

> On one hand, the definition of the state of a physical system, as ordinarily understood, claims the elimination of all external disturbances. But in that case, according to the quantum postulate, any observation will be impossible, and, above all, the concepts of space and time lose their immediate sense. On the other hand, if in order to make observation possible we permit certain interactions with suitable agencies of measurement, not belonging to the system, an unambiguous definition of the state of the system is naturally no longer possible, and there can be no question of causality in the ordinary sense of the word (PWNB, 1:54; cf. 98; NBCW, 6:399).

For example, the fixation of the position of a particle entails "a complete rupture in the causal description of its dynamical behaviour," while "the determination of its momentum always implies a gap in the knowledge of its spatial propagation" (PWNB, 1:68). This indicates that we are obliged to renounce the causal space-time mode of description which characterizes classical physics (see PWNB, 1:53, 92).

Nevertheless, Bohr by no means argues for discarding the concepts of classical theory and replacing them with new conceptual forms (see PWNB, 1:16).[9] On the contrary, he maintains, "our interpretation of the experimental material rests essentially upon the classical concepts," and thus all experience must ultimately be expressed in terms of them (PWNB, 1:53; cf. 94).[10] In other words, as he remarked

---

[8] According to Carsten Held, although Bohr "frequently equates 'causality' with 'conservation laws' and the latter with a measurement of energy and momentum," these equations "are obscure and cannot hide the fact that Bohr's original intentions are different" (1994, 882): In contrast to his mature formulations, Bohr's terminology in his Como lecture allows of the view, held by Heisenberg and von Weizsäcker, that by causality Bohr means the "deterministic description of the time evolution of the ψ-function" (von Weizsäcker 1976, 293; see Held 1994, 883). For these physicists' interpretations of complementarity, see Section 3.1 of this book. Admittedly, Bohr's phrase "the definition of the *state* of a physical system" (PWNB, 1:56, my emphasis) might seem to suggest the above Heisenbergian reading, according to which the term 'state' refers to what is expressed by the state or wave function. In my view, however, the conceptual equivalence between the claim of causality (associated with state definition) and the use of the dynamical conservation laws, while not explicitly stated by Bohr, is reasonably indicated already in the Como paper, specifically by his discussion of the connection between complementarity and the uncertainty relations to be seen below as well as by his characterization of "the conservation theorems for energy and momentum" as "complementary to the space-time description" (PWNB, 1:60, 88). See also Jammer (1974, 95).

[9] On the differences between Bohr's and Pauli's views of classical concepts, see Hendry (1984, esp. 34, 130).

[10] For parallel passages including ones from later periods, see PWMB, 1:8, 17, 2:25f., 39, 3:3, and 11. In Edward MacKinnon's account, Bohr had used the term 'classical' "in his own special sense since 1912." That is, "[b]y 'classical concepts' he meant ordinary language terms, such as 'wave' and 'particle,' which are given specialized meanings in classical physics" (MacKinnon 1985, 106).

in 1929, it is the application of classical concepts alone that "makes it possible to relate the symbolism of the quantum theory to the data of experience" (PWNB, 1:16). It should be noted that Bohr considers the concepts and terms of classical physics to be essentially bound up with "our ordinary perception" (PWNB, 1:91). By "classical concepts" he means not only the concepts specific to the discipline of classical physics, but more generally such "ideas underlying our accounts of every-day experience" as space, time, and causality (NBCW, 6:399).[11] In this sense, he emphasizes that, despite the "failure of the forms of perception adapted to our ordinary sense impressions," we cannot avoid the use of concepts associated with those forms of perception (PWNB, 1:93; cf. 19, 51).[12]

Faced with this apparent impasse – the indispensability of classical concepts despite the breakdown of the classical mode of description – we have no other choice than to impose a certain restriction on the application of classical concepts (see PWNB, 1:16). Bohr accordingly proposes that we conceive the above conflict between space-time coordination and the claim of causality as a "complementary" relationship.

> The very nature of the quantum theory thus forces us to regard the space-time co-ordination and the claim of causality, the union of which characterizes the classical theories, as complementary but exclusive features of the description, symbolizing the idealization of observation and definition respectively (PWNB, 1:54f.; cf. 95; NBCW, 6:400).[13]

In Bohr's view, this complementary relation "appears as an inevitable consequence of the contrast between the quantum postulate and the distinction between object and agency of measurement" (PWNB, 1:68).

As we will see in Chapter 3, Bohr's concept of complementarity has often been misleadingly characterized as a relation between particle and wave, or between position and momentum. Yet, as some commentators point out, in his Como paper, while using the term 'complementarity' or 'complementary' dozens of times, Bohr speaks only once of the "complementary pictures" of particles and waves, and never of the complementarity of position and momentum (PWNB, 1:56).[14] As can be readily seen from the above formulation, his concept of complementarity primarily refers

---

[11] Using quasi-Kantian terms, but in a relative deviation from Kant's position, Bohr maintains that not only space and time, but also causality "may be considered as a mode of perception by which we reduce our sense impressions to order" (PWNB, 1:116f.). For prior studies on the possible Kantian background of Bohr's complementarity, see Section 3.4.

[12] In Bohr's account, "all new experience makes its appearance within the frame of our customary points of view and forms of perception," and "we can by no means dispense with those forms of perception which colour our whole language and in terms of which all experience must ultimately be expressed" (PWNB, 1:1, 5).

[13] Although, in the text of the Como paper, this sentence directly follows the last indented quotation, I have reconstructed Bohr's reasoning linking the two by the use of other passages from the same lecture as well as from other relevant works.

[14] See Meyer-Abich (1965, 152) and Fujita (1991, 66). Although later, in his 1935 response to EPR, Bohr would speak of "complementary physical quantities" of position and momentum (PWNB, 4:80; see Folse 1985, 269), the fact remains that he only rarely does so.

to the relation between space-time coordination and the claim of causality,[15] or, in more general terms, between observation and the definition of the state (see PWNB, 1:73, 77, 87, 2:21).

This is not to say, however, that the wave-particle relation is in any way unimportant for Bohr. On the contrary, he starts from the above sense of complementarity to account for what is known as the wave-particle duality,[16] or "the paradoxical character of the problem of the nature of light and of material particles" (PWNB, 1:60; cf. 16). As regards light, it is well known that "its propagation in space and time is adequately expressed by the electromagnetic theory," and in particular that "the interference phenomena [...] are completely governed by the wave theory superposition principle" (PWNB, 1:55). According to the quantum postulate, however, "in attempting to trace the laws of the time-spatial propagation of light [...], we are confined to statistical considerations" (PWNB, 1:56). On the other hand, "the conservation of energy and momentum during the interaction between radiation and matter, as is evident in the photo-electric and Compton effect [*sic*], finds its adequate expression just in the light quantum idea put forward by Einstein" (PWNB, 1:55).[17] As for "the nature of the constituents of matter," a parallel relation holds between the conventional particle picture and the picture of the de Broglie wave. It is thus in terms of the complementarity of space-time coordination and the claim of causality that Bohr accounts for the "complementary pictures," the particle and wave pictures, of both light and matter (PWNB, 1:56; cf. 75).

According to Bohr, the above idea of complementarity has its quantitative expression in Heisenberg's relations of "reciprocal uncertainty" (PWNB, 1:57).[18] As we saw in Chapter 1, Heisenberg had formulated a set of relations for the maximum precision with which "the space-time co-ordinates and momentum-energy components of a particle" (PWNB, 1:62) or, more generally, "two canonically conjugated variables" can be simultaneously measured (PWNB, 1:73). In his Como lecture, however, Bohr not only refers to Heisenberg's results, but derives himself the uncertainty relations in a different manner, with the aid of the idea of "wave-groups" or wave packets (cf. NBCW, 6:419f.; see Jammer 1974, 92f.). He starts from Planck's and de Broglie's equations ($E = h\nu, p\lambda = h$), in which the wave and particle notions of light as well as matter "enter in sharp contrast" (PWNB, 1:58). In Bohr's account, the association of a particle with a wave packet indicates "the complementary

---

[15] Dugald Murdoch designates complementarity in this sense as "kinematic-dynamic complementarity" in distinction to "wave-particle complementarity" (1987, 58).

[16] As Sandro Petruccioli points out, however, Bohr does not accept wave-particle *dualism* in the sense of the existence of two opposing attributes of physical reality. He rather holds that electrons, for example, "are neither waves nor particles," but objects that accept a complementary mode of description (Petruccioli 1993, 173).

[17] According to Murdoch, however, this mode of correlation, supposed by Bohr, between the complementarity of space-time and causality and wave-particle complementarity does not invariably hold, and the two senses of complementarity are "logically independent notions" (1987, 67; cf. Folse 1985, 120; PWNB, 4:4f.).

[18] From the late 1930s onward, Bohr would prefer the term 'indeterminacy' to 'uncertainty' to designate Heisenberg's relations (e.g. PWNB, 2:39; cf. 4:3).

character of the description," because "the use of wave-groups is necessarily accompanied by a lack of sharpness in the definition of period and wave-length, and hence also in the definition of the corresponding energy and momentum" as given by the above equations (PWNB, 1:59). A quantitative expression of this "lack of sharpness" leads to the uncertainty relations that hold "between the maximum sharpness of definition of the space-time and energy-momentum vectors." This circumstance may be regarded as "a simple symbolical expression for the complementary nature of the space-time description and the claims of causality" (PWNB, 1:60; see Jammer 1974, 93).

This suggests how Bohr's approach to the uncertainty relations differs from Heisenberg's. First, while Heisenberg seeks to proceed with purely corpuscular considerations,[19] Bohr derives the uncertainty relations in a manner that from the start makes manifest the indispensability of both wave and particle notions.[20] Second, while Heisenberg at the time leans toward the positivist idea of reducing the possibility of definition to that of observation, Bohr sets "the possibilities of definition and observation" precisely in a complementary relationship (PWNB, 55; cf. 63).[21] As we have seen above, instead of equating definability with observability, Bohr characterizes the complementarity of space-time coordination and the claim of causality as a relation between observation and the definition of the state.

It should be noted that Bohr does not fully define the term 'complementarity' itself, nor uses it purely unambiguously. In many popular presentations of his thought, the term 'complementary' has been taken to mean simply its literal or etymological sense, 'mutually or jointly completing.' Admittedly, in the passage quoted above from his Como paper, where it reads "complementary *but* exclusive" (my emphasis), the word 'complementary' appears to be used in a meaning close to this sense, 'jointly completing,' as against that of 'exclusive.' Similarly, as regards the wave and particle pictures, Bohr says that "we are not dealing with contradictory but with complementary pictures of the phenomena, which only together offer a natural generalization of the classical mode of description" (PWNB, 1:56). In due course, however, he extends and displaces the concept of complementarity in such a way that *the meaning of mutual exclusion constitutes part of the very meaning of complementarity*. He states in a 1929 article that the term 'complementarity' denotes "the relation of mutual exclusion characteristic of the quantum theory with regard to the application of the various classical concepts and ideas"

[19] Heisenberg privileged the particle picture over the wave picture up until 1927, but, as some commentators point out, he soon shifted to the view of "equivalence" between the two pictures, which, however, still differed from Bohr's idea of wave-particle duality (see Camilleri 2009, 76ff.).

[20] This is related to Bohr's attitude toward the two versions of quantum mechanics, matrix and wave mechanics. As we saw in Chapter 1, instead of simply siding with either of the two, he viewed both theories equally as correspondence theories. In the Como paper, characterizing his approach as "harmoniz[ing] the apparently conflicting views taken by different scientists," he notes that both matrix and wave mechanics represent "symbolic transcription[s] of the problem of motion of classical mechanics adapted to the requirements of quantum theory" (PWNB, 1:52, 75). See Beller (1999, 46, 117ff.).

[21] See Jammer (1989, 351) and Hendry (1984, 125, 129, 131).

(PWNB, 1:19; cf. 4:63). Further, he speaks of complementarity "in the sense that any given application of classical concepts precludes the simultaneous use of other classical concepts which in a different connection are equally necessary for the elucidation of the phenomena" (PWNB, 1:10). The term 'complementarity' thus comes to mean – perhaps itself complementarily – both joint completion and mutual exclusion.[22]

In Bohr's view, this idea of complementarity essentially characterizes the description of nature newly offered by quantum mechanics. It would be mistaken to believe that this mode of description "leave[s] room for a mysticism which is contrary to the spirit of natural science" (PWNB, 1:15; cf. 116). Rather, complementarity refers to "the combination of features which are united in the classical mode of description but appear separated in the quantum theory," and this combination "ultimately allows us to consider the latter as a natural generalization of the classical physical theories" (PWNB, 1:19; cf. 92, 110). In other words, it is precisely by virtue of its complementary use of classical concepts that quantum mechanics may be regarded as "a rational generalization of the causal space-time description of classical physics" (PWNB, 1:87).[23]

In this section, I have outlined Bohr's 'early' complementarity argument as far as it concerns the quantum-theoretical description of nature. He does not, however, restrict his idea of complementarity to quantum physics alone. Rather, he emphasizes an essential relevance of the same idea to various other areas of knowledge and experience. In Bohr's account, the "new knowledge" associated with the concept of complementarity has profoundly "shaken the foundations underlying the building up of concepts, on which not only the classical description of physics rests but also our ordinary mode of thinking" (PWNB, 1:101). It is in this context that he enlarges his scope to include such fields as psychology and biology, approaching them from the point of view of complementarity. In the next section, I will survey Bohr's complementarity in these various fields, which appears to me, far from being superficial applications of his ideas in quantum theory, to offer essential clues for a better understanding of the latter as well.

2.2 The 'Early' Period: Complementarity in Various Fields

It would be misleading to say that Bohr, after introducing the idea of complementarity in quantum theory, simply applied it to other fields. Rather, from his youth, under the influence of his father Christian Bohr, who was a physiologist, and his philosophical mentor Harald Høffding, Bohr had long been interested in fields such as biology and psychology, and specifically in the problem of the subject-object

[22] See Murdoch (1987, 59ff.), Faye (1991, 142), and Held (1994, 871).

[23] For parallel passages, see PWNB, 1:4, 70, 2:41, 61 and NBCW, 7:317.

relation as it was discussed in these areas (see PWNB, 2:96; NBCW, 6:xxiv).[24] These interests appear to have offered him important clues for developing his idea of complementarity in quantum theory.[25] It is thus appropriate to state that Bohr's conception of complementarity has some of its roots in his prior concern with problems in biology and psychology, while his preoccupation with complementarity in quantum theory in its turn gave him further insight into those different areas. Among the fields other than quantum theory, it is psychology and, associated with it, epistemology that Bohr first approached explicitly in terms of complementarity. Let us here, however, start with his account of biological problems, which he regards as "directly connected" with both physical science and psychology (PWNB, 1:20).[26]

In 1929, Bohr touched on "the problems regarding living organisms" in connection with "our knowledge of atomic phenomena." In his account, "the mutual action between the organisms and the external world, upon which the sense impressions depend, may [...] be so small that it approaches the quantum of action." In studying physiological phenomena, one may therefore be faced with "the limit for an unambiguous description of them with the help of our ordinary visualizable conceptions." Specifically, with regard to the problems of "the freedom and power of adaptation of the organism in its reaction to external stimuli," the same conditions should be taken into account "which determine the limitation of the causal mode of description in the case of atomic phenomena" (PWNB, 1:118f.). We can see that, in hinting at the relevance of his viewpoint of complementarity to biological problems, Bohr conceives this relevance as based on the possibility of investigating the phenomena of life in terms of *physical* and specifically quantum-theoretical ideas. He does not point to "any immediate limit for the applicability" of "physical and chemical points of view" themselves to those phenomena (PWNB, 1:118). To put it differently, he still leaves open the possibility that our knowledge of atomic phenomena may eventually provide us with "a sufficient basis for tackling the problem of living organisms" (PWNB, 1:21).[27]

A few years later – in his 1931 "Addendum" to "Introductory Survey" and his 1932 lecture "Light and Life" (PWNB, 2:3–12) – however, Bohr took a decisive step. Rather than stressing the direct relevance of quantum theory to the mechanism

[24] See David Jens Adler, "Childhood and Youth," in Rozental (1967, 11–37, on 13) and Pais (1991, 99). For an extensive study of Høffding's influence on Bohr's thought, see Faye (1991, esp. Part I).

[25] It is worth mentioning the episode that, in the late 1920s, when Bohr lectured his friend, the psychologist Rubin, on the epistemological "lesson" of quantum mechanics, the latter responded: "But Niels! You told us all of that twenty years ago!" (NBCW, 6:xxvi). See Heisenberg (1967, 107) and Faye (1991, 62).

[26] See Faye (1991, 17). From 1932 onward, Bohr discussed complementarity with regard also to the phenomena of heat, that is, the "exclusive relation" between such thermodynamic concepts as temperature and "a detailed description of the behaviour of the atoms in the bodies concerned" (NBCW, 6:400; cf. PWNB, 2:97).

[27] See Paul Hoyningen-Huene, "Niels Bohr's Argument for the Irreducibility of Biology to Physics," in Faye and Folse (1994, 231–55, on 231f.).

of living organisms (see PWNB, 2:8), he now points out that "there is set a fundamental limit to the analysis of the phenomena of life in terms of physical concepts" (PWNB, 1:22). Instead of reducing the phenomena of life to quantum-physical processes, he critically examines the very applicability of physical concepts to those phenomena. This does not mean, however, that he advocates restoring an ontological principle specific to life, traditionally referred to as vital force, which, in his view, "can hardly be given an unambiguous expression." Rather, "if [...] we were able to push the analysis of the mechanism of living organisms as far as that of atomic phenomena, we should not expect to find any features foreign to inorganic matter" (PWNB, 2:9). What is crucial, however, is that such a physical analysis involves an unavoidable interaction with the organism in question. Particularly, "the interference necessitated by an observation which would be as complete as possible from the point of view of the atomic theory would cause the death of the organism" (PWNB, 1:22). We would "doubtless kill an animal," for example, "if we tried to carry the investigation of its organs so far that we could tell the part played by the single atoms in vital functions." For this reason, "the necessity of keeping the object of investigation alive imposes a restriction" on the physical analysis (PWNB, 2:9; cf. 20).

Here lies, argues Bohr, an essential analogy with the situation facing us in quantum theory. As he puts it,

> [...] the very existence of life must in biology be considered as an elementary fact, just as in atomic physics the existence of the quantum of action has to be taken as a basic fact that cannot be derived from ordinary mechanical physics. Indeed, the essential non-analyzability of atomic stability in mechanical terms presents a close analogy to the impossibility of a physical or chemical explanation of the peculiar functions characteristic of life (PWNB, 2:9; cf. 1:23, 3:26).

This suggests that the idea of complementarity, introduced in quantum theory, becomes also relevant to the field of biology. Bohr thus states that "the strict application of those concepts which are adapted to our description of inanimate nature might stand in a relationship of exclusion to the consideration of the laws of the phenomena of life" (PWNB, 1:22f.). In other words, the contrast "between such typical aspects of life as the self-preservation and the self-generation of individuals, on the one hand, and the subdivision necessary for any physical analysis on the other hand" may be regarded as a relation of complementarity. According to Bohr, an important implication of this "feature of complementarity" is that "the concept of purpose, which is foreign to mechanical analysis, finds a certain field of application in biology" (PWNB, 2:10).[28]

This complementarity argument by Bohr with regard to biological problems is, as noted earlier, chronologically preceded by his account of a comparable state of affairs in the fields of psychology and epistemology. Already in his Como paper,

---

[28] As he himself later notes, this insight of Bohr into an irreducible role of the notion of purpose in biology was originally inspired by the views of his father, the physiologist Christian Bohr (PWNB, 2:96). See Folse (1985, 45).

Bohr suggested a possible extension of the idea of complementarity in that direction: "[...] the idea of complementarity is suited to characterize the situation which bears a deep-going analogy to the general difficulty in the formation of human ideas, inherent in the distinction between subject and object" (PWNB, 1:91).[29] He set out to elaborate this view from 1929 onward, by stating that "[t]he impossibility of distinguishing in our customary way between physical phenomena and their observation places us, indeed, in a position quite similar to that which is so familiar in psychology where we are continually reminded of the *difficulty of distinguishing between subject and object*" (PWNB, 1:15). This difficulty lies in the following circumstance:

> For describing our mental activity, we require, on one hand, an objectively given content to be placed in opposition to a perceiving subject, while, on the other, as is already implied in such an assertion, no sharp separation between object and subject can be maintained, since the perceiving subject also belongs to our mental content (PWNB, 1:96).

Just as an "unavoidable influence on atomic phenomena" is caused by observing them, one experiences a "change of the tinge of the psychological experiences" when one directs attention to their elements (PWNB, 1:100). It is for this reason that "[t]he necessity of taking recourse to a complementary, or reciprocal,[30] mode of description is perhaps most familiar to us from psychological problems" (PWNB, 1:96).

Bohr particularly focuses on the age-old philosophical theme of the relation between free will and causality, by pointing to the phenomena that are both "experienced as free will and analyzed in terms of causality" (PWNB, 1:24; cf. 116). He at times treats this relationship as a relation between psychical experience and its *physiological* aspect. Specifically, he argues, "the contrast between the feeling of free will, which governs the psychic life, and the apparently uninterrupted causal chain of the accompanying physiological processes" may be regarded as a relation of complementarity. For, just as any attempt at "a detailed causal tracing of atomic processes [...] involves a fundamentally uncontrollable interference with their course," so also an attempt to observe "the processes in the brain" will bring about "an essential alteration in the awareness of volition" (PWNB, 1:100f.).

Bohr does not, however, restrict causality to the physiological dimension. Rather, he claims more generally that "causality may be considered as a mode of perception by which we reduce our sense impressions to order," while "the freedom of the will

[29] This sentence was not included in Bohr's original lecture, but added afterward for the version published in *Nature* (see NBCW, 6:112, 136). Jan Faye attributes this development primarily to Høffding's influence on Bohr in their discussion during the period (Faye 1991, 60).

[30] After introducing the term 'complementarity,' Bohr replaced it by 'reciprocity' for a short time around 1929, but subsequently reverted to the former (see PWNB, 1:19). See also Murdoch (1987, 60).

is an experiential category of our psychic life" (PWNB, 1:116f.).[31] Most importantly, "introspection" of one's own psychical experience exerts an unavoidable influence on it, and this "shows a striking similarity to the conditions responsible for the failure of causality in the analysis of atomic phenomena" (PWNB, 1:23f.).[32] This state of affairs concerning introspection may be illustrated by a passage from a novel of the Danish poet and philosopher Poul Martin Møller, *The Adventures of a Danish Student*, which Bohr considers to be highly relevant. In it, one of the characters, nicknamed "the licentiate," describes his endless series of self-reflections:

> [...] I get to think of my own thoughts of the situation in which I find myself. I even think that I think of it, and divide myself into an infinite retrogressive sequence of "I's" who consider each other. I do not know at which "I" to stop as the actual, and in the moment I stop at one, there is indeed again an "I" which stops at it (cited in PWNB, 3:13; see Folse 1985, 54).[33]

This experience, continues the licentiate, makes him feel as if he were "looking down into a bottomless abyss" (cited in PWNB, 3:13).

This complementarity of psychical experience and its reflective analysis has, in Bohr's view, direct implications for epistemological problems, specifically for the problem of the "shaping of concepts" (PWNB, 1:15). Since the analysis of a concept unavoidably interferes with our state of mind, "the conscious analysis of any concept stands in a relation of exclusion to its immediate application" (PWNB, 1:96),[34] and this "complementary relationship" holds "in all domains of knowledge" (PWNB, 1:20). This leads to "the relative meaning of every concept, or rather of every word, the meaning depending upon our arbitrary choice of view point," and thus also to the "subjective character of all experience" (PWNB, 1:96, 1; cf. 97; NBCW, 6:46).

---

[31] Folse points to a "striking similarity" between this account by Bohr and Kant's account of free will (Folse 1985, 52f.; cf. Faye 1991, xiv). As we will see in Section 3.4, however, he denies an overall affinity between Bohr's thought and Kantian philosophy.

[32] Bohr would amplify this point in subsequent years – well into what I call the 'middle' period. In 1938, he argues: In introspection, "the phenomena themselves and their conscious perception" are "mutually exclusive" and "complementary" to each other, as is illustrated by the fact that "if we try to analyze our own emotions, we hardly possess them any longer" (PWNB, 2:27). Faye points to a similarity of this view to Høffding's account of the antinomy of "involuntary mental life" and "reflection" (Faye 1991, 95f.).

[33] Although this quotation is from his later work (1960), Bohr, already in a 1928 speech, referred to Møller's novel in question and its description of the "poor licentiate," which expresses "the struggle between opposites within our own mind" (NBCW, 10:234f.). According to Léon Rosenfeld (1967, 121) as well as Bohr's son Aage (PWNB, 3:vi), Bohr was, from his youth, familiar with this novel and deeply interested in its implications for the problem of self-reflection. Incidentally it is also from his early years that, in considering the above situation concerning introspection, Bohr drew an analogy with multivalued functions of complex variables. See *Archive for the History of Quantum Physics*, pp. 1f. See also Folse (1985, 51f., 176f.).

[34] Bohr would later reformulate this complementarity as the relation between "the practical use of any word and attempts at its strict definition," or between "the direct use of any word" and "an analysis of its meaning" (PWNB, 2:52, 4:91).

The idea of complementarity is more radical, however, than the mere notion of the relativity or subjectivity of experience. Complementarity in the epistemological dimension implies, according to Bohr, an essential "ambiguity in our use of language," such that "even words like 'to be' and 'to know' lose their unambiguous meaning" (PWNB, 1:19). For this reason, even in natural science, "there can be no question of a strictly self-contained field of application of the logical principles" (PWNB, 1:97; see Held 1994, 879). If we may draw on Heisenberg's recollection of his discussion with Bohr in 1933, the latter remarked that "we never know exactly what a word means," because language has a "peculiarly floating (*schwebend*) character." This applies, he continued, not only to "the language of the poets," but also to "ordinary language," and further "to some extent to the language of natural science" (Heisenberg 1969, 161/134f., trans. mod.). This is why, in natural science, we can at best approach the ideal of logically strict formulation, but "definitely cannot attain it." Heisenberg further quotes Bohr as saying: "In religion, from the outset we abandon the idea of giving an unambiguous meaning to words, while in natural science one starts from the hope – or from the *illusion* – that it could some day in much later times be possible to give an unambiguous meaning to words" (1969, 162/135f., trans. mod., my emphasis).[35]

Here I wish briefly to touch on Bohr's notion of "deep truth," which he did not thematically expound in his writings, but is reported to have embraced and often privately expressed in connection with epistemological problems. According to his son Hans Bohr, "[o]ne of the favorite maxims of my father was the distinction between the two sorts of truths, profound truths recognized by the fact that the opposite is also a profound truth, in contrast to trivialities where opposites are obviously absurd."[36] Although it is not immediately clear how this conception of truth is linked with the above idea of complementarity in epistemology, we can find a minimal clue in the following account by Paul Dirac. According to Dirac, Bohr held that the ambiguity in the meaning of words, as mentioned above, may also "govern the truth or falsity of a statement." Since "a statement of the highest wisdom," in particular, "necessarily involves words whose meaning cannot be defined unambiguously," its truth "is only relative to a suitable meaning for the ambiguous words in it, with the consequence that the converse statement also has validity and is also wisdom." Bohr illustrated this with the statement "There is God" and its converse "There is no God," both of which he designated as "great wisdom and truth."[37]

In the previous and the present sections, I have outlined Bohr's complementarity argument with regard to quantum theory as well as other fields of knowledge

---

[35] This remark by Bohr appears in the context of a critique of positivism. In questioning the possibility of attaining unambiguous meanings, he is quoted as saying: "I object to positivism not because I would be less skeptical [than positivists] in this area, but because, on the contrary, I am afraid that, on principle, things could not at all be much better in natural science" than in religion (Heisenberg 1969, 162/135f., trans. mod.).

[36] Hans Bohr, "My Father," in Rozental (1967, 325–39, on 328).

[37] Paul A. M. Dirac, "The Versatility of Niels Bohr," in Rozental (1967, 306–09, on 309).

and experience including epistemology. As can be seen from the above discussion, complementarity in epistemology does not simply stand side by side with that in other fields, but is concerned with the logical relations common to "all domains of knowledge" (PWNB, 1:20; cf. 2:2, 27, 3:7).[38] In other words, Bohr seeks to connect the situations of complementarity in various fields in terms of "the relation between subject and object which forms the core of the problem of knowledge" (PWNB, 1:117). He often expresses this common relationship by the metaphorical dictum that "we are both onlookers and actors in the great drama of existence" (PWNB, 1:119; cf. 2:20, 63).[39] As regards quantum theory, we can see that space-time coordination, which involves an interaction between the object and agency of observation, has the character of being an 'actor,' while the claim of causality, which excludes such an interaction, has the mode of being a 'spectator.' To take psychological problems as another example, it is as an 'actor' that one has psychical experience, while it is as a 'spectator' that one reflects on that experience. In this manner, Bohr conceives complementarity in different fields generally as a relation of contrast between the roles of 'actors' and 'spectators.'[40] It will turn out, however, that the situations of complementarity in different fields cannot be so easily interconnected as it might at first sight appear. Later in the present study, in Chapter 4, I will address this issue by rendering the analogy between quantum theory and other fields more explicit than Bohr himself does.

2.3 The Debate with EPR and the 'Middle' Period

Bohr's complementarity, as it was formulated in the 'early' period, did not remain unchanged throughout his life. Commentators are more or less in agreement that, starting in the mid-1930s, his complementarity argument did undergo alterations *at least* on the terminological level. The question remains unsettled, however, whether this development constitutes a *philosophical* transformation of Bohr's thought. As we will see in Chapter 3, some commentators, including myself, speak of significant shifts in his philosophical orientation, while others perceive nothing more than a terminological refinement (see Faye and Folse 1998, 2). In the remaining sections of this chapter, however, I largely suspend this interpretive question, and restrict myself for the moment to those changes which explicitly appear in his texts, that is, visible revisions and modifications in terminology and other aspects of argumentation.

[38] We may also refer to the following expression by Bohr, though from the early 'middle' period: the "*epistemological* lesson which the opening of quite new realms of physical research has given us in the latest years" (PWNB, 2:24, my emphasis; cf. 3:12).

[39] In Jørgen Kalckar's account, Bohr may have picked up this metaphor also from Møller's work mentioned above, *Adventures of a Danish Student*. In it, with regard to his endless introspection, the licentiate says that "the spectator anew becomes an actor" (cited in NBCW, 6:xxii). See also Favrholdt (1992, 53).

[40] In a 1928 speech, as regards one's intellectual life in general, Bohr remarks that "we are not just observers, but are participants ourselves" (NBCW, 10:234).

The first major change of this kind occurred in Bohr's 1935 debate on quantum mechanics with Albert Einstein and his collaborators.[41] In this year, Einstein, Podolsky and Rosen (commonly abbreviated as EPR) jointly presented a paper titled "Can Quantum-Mechanical Description of Physical Reality Be Considered Complete?," which purports to demonstrate the incompleteness of quantum mechanics.[42] Their argument may be briefly outlined as follows.[43] To show that there are elements of reality which cannot be described by quantum mechanics, the authors begin by formulating the following criterion of reality: "If, without in any way disturbing a system, we can predict with certainty (i.e., with probability equal to unity) the value of a physical quantity, then there exists an element of physical reality corresponding to this physical quantity" (Einstein, Podolsky, and Rosen 1935, 777).[44] They proceed to introduce two physical systems, I and II, which, having previously interacted for some time, are no longer in interaction with each other. Suppose, the authors continue, one knows the state of the combined system I + II, and measures a physical quantity of system I alone and thereby indirectly makes a prediction on system II. While the measurement on system I disturbs that system, they argue that "since [. . .] the two systems no longer interact, no real change can take place in the second system in consequence of anything that may be done to the first system" (1935, 779). According to the uncertainty principle, one can measure precisely either the position or the momentum of system I. By measuring the exact position of system I, one can "predict with certainty" the position of system II "without in any way disturbing" that system. The same holds for the momentum

---

[41] The debate between Einstein and Bohr on quantum theory had begun in the 1920s, developed notably on the occasions of two Solvey Cogresses, the Fifth (1927) and the Seventh (1930). This development was reviewed by Bohr himself in 1949 (PWNB, 32–66). See also Jammer's account in 1974, 108–58.

[42] As Arthur Fine points out, "[f]or reasons of language" this EPR paper was not actually written by Einstein, but by Boris Podolsky. Rather unsatisfied with the outcome, Einstein remarked that "the essential thing was, so to speak, smothered by the formalism (*Gelehrsamkeit*)" (cited in Fine 1986, 35).

[43] According to Fine, the EPR argument in the original text may be schematically reconstructed as follows:

$$
\begin{array}{l}
(\mathrm{INC}) \vee (\mathrm{NSV}), \\
\sim (\mathrm{INC}) \rightarrow \sim (\mathrm{NSV}). \\
\therefore (\mathrm{INC})
\end{array}
$$

Here (INC) stands for the assertion of the incompleteness of quantum mechanical description, and (NSV) for the statement that "observables represented by noncommuting operators cannot have simultaneous reality" (Fine 1986, 32; see also Jammer 1974, 181ff.; Murdoch 1987, 165ff.; Kaiser 1992, 229). In Fine's view, however, the authors actually "establish the conditional ~(INC) → ~(NSV) simply by deriving the consequent ~(NSV)" without using the assumption of completeness. For this reason, the above logical structure of their argument "seems strangely complex" (1986, 33f.). In view of this circumstance, with many commentators I represent the EPR argument in a reasonably simplified form. See, for example, Selleri (1990, 109–15).

[44] In Fine's account, this EPR criterion of reality cannot be "a principle that Einstein put much stock by," for "it never appears in any of his own published expositions of the EPR situation" (1986, 62).

as well. Applying the above criterion of reality, EPR therefore assert that system II must have simultaneously definite position and momentum, despite the fact that quantum-mechanical description allows no object to have simultaneously definite position and momentum. The authors thus conclude that "the quantum-mechanical description of physical reality [...] is not complete" (1935, 780).

This EPR argument exerted a serious impact on Bohr,[45] urging him to reexamine and reformulate his ideas concerning observational interaction. Later in the same year, Bohr responded to the EPR challenge in an article with the same title.[46] In it, he points out that EPR's criterion of reality contains "an essential ambiguity" (PWNB, 4:75) with regard to the expression "without in any way disturbing a system." To explicate this, Bohr begins by supposing a diffraction experiment using a diaphragm with a single slit through which a particle passes (see PWNB, 4:76). As is well known, depending on whether the diaphragm is rigidly or loosely attached to the rest of the apparatus, one can measure either the position or the momentum of the particle (see Murdoch 1987, 168). As Bohr puts it, here "we are not dealing with an incomplete description characterized by the arbitrary picking out of different elements of physical reality [...], but with a rational discrimination between essentially different experimental arrangements and procedures which are suited either for an unambiguous use of the idea of space location or for a legitimate application of the conservation theorem of momentum." To put it differently, "we have in each experimental arrangement [...] not merely to do with an ignorance of the value of certain physical quantities, but with the impossibility of defining these quantities in an unambiguous way" (PWNB, 4:78).

Now, contends Bohr, this "appl[ies] equally well to the special problem treated by Einstein, Podolsky and Rosen, [...] which does not actually involve any greater intricacies." For the EPR situation "may be reproduced, at least in principle, by a simple experimental arrangement, comprising a rigid diaphragm with two parallel slits, which are very narrow compared with their separation, and through each of which one particle [...] passes independently of the other" (PWNB, 4:78, see Jammer 1974, 195). If we choose to "measure the position of one of the particles," we allow "an essentially uncontrollable momentum to pass from the first particle into the [...] support," and thereby "cut ourselves off from any future possibility of applying the law of conservation of momentum to the system" and therefore lose "our only basis for an unambiguous application of the idea of momentum in predictions regarding the behaviour of the second particle." Conversely, continues Bohr, "if we choose to measure the momentum of one of the particles, we lose through the uncontrollable displacement [...] any possibility of deducing [...] the position of the diaphragm relative to the rest of the apparatus, and thus have no basis whatever

[45] As Rosenfeld recalls, "This onslaught came down on us as a bolt from the blue. Its effect on Bohr was remarkable" (1967, 128).

[46] Niels Bohr, "Can Quantum-Mechanical Description of Physical Reality be Considered Complete?," *Physical Review* 48 (1935): 696–702; reprinted in NBPW, 4:73–82. Just as in the case of the EPR argument, my outline of Bohr's reply here leaves aside various interpretive issues concerning the debate. See, for instance, Beller (1999, 145–67).

for predictions regarding the location of the other particle" (PWNB, 4:79f.). This consideration leads Bohr to claim:

> Of course there is in a case like that just considered no question of a mechanical disturbance of the system under investigation during the last critical stage of the measuring procedure. But even at this stage there is essentially the question of *an influence on the very conditions which define the possible types of predictions regarding the future behavior of the system* (PWNB, 4:80; see Murdoch 1987, 170).

Since, in Bohr's view, these conditions are constitutive of the description of what may be called physical reality, EPR's argumentation "does not justify their conclusion that quantum-mechanical description is essentially incomplete." It is precisely this new situation as regards the description of physical phenomena that "the notion of complementarity aims at characterizing" (PWNB, 4:80). Bohr concludes by noting that "this new feature of natural philosophy means a radical revision of our attitude as regards physical reality" (PWNB, 4:82; cf. 75, 2:61).[47]

This response by Bohr to the EPR argument appears to have opened a new phase of the development of his thought, in which he sought further to reorganize his complementarity argument.[48] In the rest of this section, I will sketch the development during the 'middle' period, stretching from 1935 until around 1950 – including, in particular, his extensive 1949 article "Discussion with Einstein on Epistemological Problems in Atomic Physics" (PWNB, 2:32–66). As it turns out, during this period he made a series of changes in his terminology and argumentation with regard to quantum theory, which may be summarized as follows.

(1) Let us first look at Bohr's new point concerning the notion of "disturbance," which has drawn much attention among commentators. In his above response to EPR, while stressing an "influence" on the predictability of phenomena, which is different from a "mechanical disturbance," he did not explicitly abandon the concept of "disturbance" itself. From the late thirties on, however, Bohr not only refrained from, but explicitly rejected the use of phrases such as "disturbing of phenomena by observation" (PWNB, 2:63f.; cf. 73). As he put it in 1938, "[s]peaking, as is often done, of disturbing a phenomenon by observation, or even of creating physical attributes to objects by measuring processes, is, in fact, liable to be confusing." For all such expressions "imply a departure from basic conventions of language which [...] can never be unambiguous" (NBCW, 7:316).[49]

The target of Bohr's critique here is not simply the use of such *words* as 'to disturb' and 'to create.' Rather, he argues generally against speaking as if the observer "influence[d] the events which may appear under the conditions he has arranged,"

---

[47] This is worth comparing with his later, more 'modest' expression (from 1956): "a revision of the foundation for the unambiguous application of some of our most elementary concepts" (PWNB, 4:171).

[48] During World War II, Bohr's writings on complementarity and related subjects are naturally rather scarce. In 1943, while Denmark was occupied by Nazi Germany, he fled the country through Sweden to the United States, where he stayed until returning to Denmark in 1945. For a biographical account of Bohr's life during the war, including his role in the atomic weapons program, see Pais (1991, 479–504).

[49] For parallel passages, see NBCW, 7:335; PWNB, 2:63f., 73, and 3:5. See also Pais (1991, 432).

because it is "hardly compatible with common language and practical definition" (PWNB, 2:51, 63f.). Linked with this, he also draws a clear conceptual line (with regard to phenomena) between observation and experimental arrangement. Admittedly, his remark in 1938, for instance, that "the interaction between the object and the measuring instruments will have an essential influence on the phenomenon itself" (PWNB, 4:100) was still in keeping with his earlier notion of the "influence on atomic phenomena caused by observing them" (PWNB, 1:100). In due course, however, correlatively with a redefinition of the phenomenon to be seen below, he came to hold that the phenomenon is *not* influenced by observation or observational interaction, but only conditioned by the experimental arrangement. To be sure, this is not to deny that observation involves an "uncontrollable interaction between the objects and measuring instruments" (PWNB, 2:41; cf. 74). Yet, this interaction is conceptually separated from the conditioning of the phenomenon by the experimental arrangement. In other words, Bohr separates the experimental arrangement from the overall observational influence, reserving the term 'interaction' for the effect that occurs *under* the given experimental arrangement.

(2) Associated with the above change, Bohr revises the use of the term 'phenomenon' itself. As can be seen from our survey in Section 2.1, in his early formulation of complementarity the term 'phenomenon' had partly overlapped with that of the object of observation, while also referring to what appears as the effect of observation. From the late 1930s on, however, he called the former the 'atomic object' and only the latter the phenomenon, thus sharply distinguishing the two terms. In his new terminology, the phenomenon is no longer that which the object *becomes* when it is observed and thereby subjected to an interaction with the measuring instrument (see Faye 1991, 192). Rather, Bohr defines the term 'phenomenon' as "the effect observed under given experimental conditions" (NBCW, 7:316). In other words, he proposes that one use the term "exclusively to refer to the observations under specified circumstances, including an account of the whole experimental arrangement" (PWNB, 2:64; cf. 73).[50] This implies that the phenomenon is not something that interacts with the measuring instrument, but rather contains in itself such an interaction.[51]

(3) Linked with this new concept of the phenomenon, Bohr redefines the very term 'complementarity.' As he puts it in a 1938 lecture,

> [...] phenomena defined by different concepts, corresponding to mutually exclusive experimental arrangements, can unambiguously be regarded as complementary aspects of the whole evidence concerning the objects under investigation (NBCW, 7:316f.).[52]

---

[50] See MacKinnon (1985, 119) and Pais (1991, 432). In Bohr's account, with this new terminology, "the observational problem is free of any special intricacy since, in actual experiments, all observations are expressed by unambiguous statements referring, for instance, to the registration of the point at which an electron arrives at a photographic plate" (PWNB, 2:64; cf. 51).

[51] Bohr would make this more explicit in the 'late' period by saying, for example, that "the interaction between the objects and the measuring instruments [...] forms an integral part of the phenomena" (PWNB, 2:72; cf. 74, 98, 3:4).

[52] For parallel passages, see NBCW, 7:419; PWNB, 2:19, 40, 74, 90, 99, 3:5, 12, 25, and 92. This differs in terminology from the following remark by Bohr in 1937, for instance: "the aspects of

That is, Bohr reconceptualizes complementarity as a relation of different phenomena – phenomena in the above redefined sense – that appear under mutually exclusive experimental conditions. Bohr applies the term 'complementary' also to the pieces of information on, or kinds of "evidence" of, such complementary phenomena (PWNB, 2:40; cf. 4:3). That is: "Information regarding the behaviour of an atomic object obtained under definite experimental conditions may [...] be adequately characterized as *complementary* to any information about the same object obtained by some other experimental arrangement excluding the fulfilment of the first conditions" (PWNB, 2:26; cf. 4:84f.).[53] Here the word 'object' appears in the singular, and this makes manifest Bohr's notion that complementarity is a relation between two phenomena (or corresponding two kinds of information) concerning *one* and the same object.[54]

With the series of terminological and conceptual changes outlined so far in mind, we can briefly summarize Bohr's 'middle'-period complementarity argument with regard to quantum theory as follows. As in the 'early' period, Bohr starts by pointing to "the peculiar feature of indivisibility, or 'individuality,'" of quantum phenomena (PWNB, 2:34),[55] which implies that any observation of atomic processes involves an "unavoidable interaction between the objects and the measuring instruments" (PWNB, 2:25; cf. 19, 41). In fact, he continues, "any attempt of subdividing the phenomena" – specifically, any attempt to analyze the observational interaction in such a way as to make a sharp object/instrument distinction – "will demand a change in the experimental arrangement introducing new possibilities of interaction between objects and measuring instruments which in principle cannot be controlled" (PWNB, 2:40).[56] Owing to this "uncontrollable" character of the interaction, "no sharp separation can be made between an independent behaviour of the objects and their interaction with the measuring instruments" (PWNB, 2:41, 52),[57] so that "an essential element of ambiguity is involved in ascribing conventional physical attributes to atomic objects" (PWNB, 2:40).[58] This constitutes "an

quantum phenomena revealed by experience obtained under such mutually exclusive conditions must thus be considered complementary" (PWNB, 2:19, my emphasis). This statement, in which it is not phenomena themselves, but their aspects that are called complementary, indicates an earlier stage of his conception.

[53] Bohr would later speak of the complementary relation between different "experiences" as well (PWNB, 2:74). It is also noteworthy that, as Faye and Folse point out, wave-particle complementarity disappeared from Bohr's complementarity argument in and after the 1940s (Faye and Folse 1998, 3). See also Held (1994, 872, 880ff.).

[54] See Held (1994, 886). Although, already in 1929, Bohr had spoken of "one and the same object" in the statement that "a complete elucidation of one and the same object may require diverse points of view which defy a unique description" (PWNB, 1:96), he had not formulated a relation between the single object and two *phenomena*.

[55] For parallel expressions, see PWNB, 2:17, 24, 33, 39, 58, and 62. In the 'late' period, Bohr would largely replace the term 'individuality' by "wholeness" (PWNB, 2:72, 85, 3:2).

[56] For parallel passages, see PWNB, 4:85, 101; and NBCW, 7:311f. For partly corresponding remarks from the 'late' period, see PWNB, 2:72, 90, and 99. See also Murdoch (1987, 91).

[57] For parallel passages, see PWNB, 2:25, 30, 39f., 47, 98, 4:86f., and 100.

[58] For parallel passages, see PWNB, 2:51, 61, 4:86, 174; and NBCW, 7:335.

epistemological problem quite new in natural philosophy" (PWNB, 2:25). Bohr stresses, on the other hand, that "however far the phenomena transcend the scope of classical physical explanation, the account of all evidence must be expressed in classical terms" (PWNB, 2:39). For this reason, evidence obtained under mutually exclusive experimental conditions "cannot be comprehended within a single picture, but must be regarded as *complementary* in the sense that only the totality of the phenomena exhaust the possible information about the objects" (PWNB, 2:40).[59] Specifically, the phenomena that appear under experimental arrangements specified in terms of "space-time concepts" are complementary to the phenomena that appear under experimental conditions determined by the "dynamical conservation laws" (PWNB, 2:40f.; cf. 58). This, in outline, is how Bohr presented his renewed complementarity argument in the 'middle' period.[60]

It is worth noting that, from the beginning of the development of his complementarity idea, Bohr was faced with certain undesirable reactions from among physicists and some researchers in other fields (see PWNB, 1:15, 116). Specifically, some of his expressions evoked "in many minds [...] the impression of an underlying mysticism foreign to the spirit of science." It appeared to him that such "misunderstandings" (PWNB, 2:63) lay at the root, in particular, of Einstein's persistent objection to his viewpoint of complementarity.[61] For this reason, he had to stress repeatedly that complementarity "in no way impl[ies] acceptance [...] of any mysticism" (PWNB, 2:20),[62] and, on other occasions, that his epistemological analogy contains nothing like "purely metaphysical speculations" (PWNB, 4:90).[63] Bohr's above effort to reformulate his complementarity argument during the 'middle' period may be viewed as a response to such "misunderstandings" and an endeavor to clear them up. As we have seen above, Bohr's conceptual development due to this effort culminated with his article "Discussion with Einstein." This work stands

[59] In the 'middle' period, Bohr often characterizes his idea of complementarity in terms of the conceptual pair of "analysis" and "synthesis" (PWNB, 2:18, 33, 47, 52, 58, 62f., 65, 68, 4:84, 88; NBCW, 7:335). He does not, however, elaborate the meanings of these terms, nor does he always use them in a clear-cut manner. On the one hand, he speaks of "analysis" and "synthesis" as corresponding to the two meanings of complementarity, "mutual exclusion" and joint completion, respectively (PWNB, 4:125). On the other, he also uses the term 'analysis' when characterizing the "individuality of quantum effects" as the impossibility of analysis (PWNB, 2:62; cf. 19, 4:128, 134; NBCW, 10:59). For accounts of Bohr's ideas of analysis and synthesis, see Chevalley (1995, 20) and Falkenburg (1998, 105ff.).

[60] For Bohr's 'middle'-period complementarity argument with regard to biology and psychology, see PWNB, 2:20–22, 27ff., and 4:88–91. See also Folse (1985, 175ff.). Because of a relative scarcity of his relevant remarks, however, it seems to be difficult to approach thematically this part of his work in connection with, and in distinction from, his earlier or later thought concerning the same fields.

[61] Bohr admits that, at a 1936 conference, in particular, despite his efforts to clear up such misunderstandings, he "had in this respect only little success in convincing my listeners" (PWNB, 2:63).

[62] For parallel passages, PWNB, 2:20, 27, 91, and 4:83.

[63] In this connection, commentators have also suggested a possible role of the logical positivist Otto Neurath's critique of Bohr's 'metaphysical' mode of expression. See Section 3.4.

at the same time, however, at the starting point of a new stage in the development of his thought. In the next section, I will accordingly proceed to the 'late' period – the final phase of his intellectual career – in which he further reoriented his idea of complementarity.

2.4 Complementarity in the 'Late' Period

From around 1950 until the end of his life, Bohr might, on the face of it, seem to have just repeatedly stated the views he had already established. While, however, taking as a point of departure his complementarity argument reformulated in the 'middle' period, his further presentations of complementarity in the 'late' period prove to have some new or newly manifested characteristics. Let us begin with his changed attitude toward the notion of paradox or paradoxical truth. As mentioned earlier, distinguishing between two kinds of truth, Bohr had long conceived "deep" or paradoxical truth – the truth whose negation is also a truth – as closely associated with his idea of complementarity.[64] Toward the end of his 1949 article "Discussion with Einstein," however, he wrote as follows: To be sure, "deep truth" prevails in the intermediate stage of the development of physical theories (PWNB, 2:66). Now, however, that "a veritable crisis in physical science has been overcome," we are nearing the goal "where logical order to a large extent allows us to *avoid* deep truth" (PWNB, 2:65f., my emphasis). Starting with this move, in the 'late' period Bohr views his idea of complementarity no longer as a paradoxical deep truth, but as a path toward "plain truth" by removing apparent paradoxes (see PWNB, 4:162).[65]

This avoidance of deep truth appears to be connected with Bohr's increasing emphasis on the notions of the "harmony" and "unity" of human knowledge (PWNB, 3:22).[66] To be sure, Bohr had from the beginning conceived his complementarity argument as serving to "harmonize the apparently conflicting views" (PWNB, 1:52; cf. 24, 101, 2:11). Yet his notion of harmony in the 'early' period

[64] In his Como paper, for example, Bohr noted that Heisenberg's work "elucidate[s] many paradoxes appearing in the application of the quantum postulate" (PWNB, 1:73). Also in 1927, he wrote that quantum paradoxes "are deeply rooted in nature." A letter from Bohr to Bidhubhusan Ray, January 22, 1927, cited in Bohr (1985, 138). See also Murdoch (1987, 46).

[65] Linked with this avoidance of paradoxical truth, Bohr stresses that quantum mechanics "fulfil[s] all demands on rational explanation with respect to consistency and completeness" (PWNB, 3:6). To be sure, his characterization of quantum mechanics as "a rational generalization of the classical theories" had itself been a constant claim since the 'early' period (PWNB, 1:70). While, however, he had earlier simultaneously pointed to the "irrationality" that the quantum postulate brought into the description of nature, he no longer speaks of irrationality – not even with the qualifying phrase "from the point of view of the classical theories" (PWNB, 1:7) – but exclusively emphasizes the rationality of quantum-mechanical description (see PWNB, 2:90, 100, 3:2).

[66] See also PWNB, 2:20, 33, 63, 65, 68, 3:6, 10, 14, 4:91, and 190. Bohr remarked in 1956 that "the relationship between cultures may [...] be regarded as complementary" and, in particular, that contacts between them "may even lead to *a common culture* with a more embracing outlook" (PWNB, 4:178f., my emphasis).

had at times assumed a paradoxical character, as illustrated by the following remark from 1928: Harmony in human thinking "can [...] only be dimly perceived, never grasped; at any attempt to do so, it slips according to its very nature through our fingers" (NBCW, 10:235). By contrast, in his 1954 conference address "Unity of Knowledge" (PWNB, 2:67–82), he characterizes his thought centering on complementarity as "the endeavour to achieve a harmonious comprehension of ever wider aspects of our situation, recognizing that [...] any apparent disharmony can be removed only by an appropriate widening of the conceptual framework" (PWNB, 2:82). In general, while his 'early' notion of harmony had implied a paradoxical correlation of conflicting moments, that of the 'late' period means a plain, harmonious unity free of paradoxes.

Associated with the tendencies mentioned so far, what is crucial to Bohr's complementarity argument in the 'late' period is his renewed account of the objectivity and unambiguity of the description of nature. He seeks to establish the objectivity of quantum-theoretical description by focusing on the problem of language and communication or, in other words, through reflection on "the character and scope of our means of communication" on which "every analysis of the conditions of human knowledge must rest" (PWNB, 2:88).[67] Bohr closely associates, or rather virtually equates, the concept of objectivity with that of "unambiguous communication." That is, by "objective description" he means the "communication of information" in which "no ambiguity is involved" (PWNB, 2:67, 3:3). According to Bohr, one must account for experience "in a manner independent of individual subjective judgment and therefore objective in the sense that it can be unambiguously communicated" (PWNB, 3:10).

Bohr argues that our "common human language," including the terminology of classical physics, serves as the medium of such unambiguous communication (PWNB, 3:10; cf. 1; NBCW, 7:316). To be sure, already in the "early" period he had maintained that "only with the help of classical ideas is it possible to ascribe an unambiguous meaning to the results of observation" (PWNB, 1:17). He had not, however, considered this unambiguous meaning to be a guarantee for the overall unambiguity of the description of nature. In fact, as we have seen earlier, Bohr had also stressed that, in all domains of knowledge, complementarity implies an "ambiguity in our use of language," such that "even words like 'to be' and 'to know' lose their unambiguous meaning" (PWNB, 1:19). In the 'late' period, however, he no longer sees such an insurmountable limit to the unambiguity of language. Rather, he now holds that our common language, which is characterized by its sharp distinction between subject and object, constitutes a solid basis for unambiguous description and communication (see PWNB, 3:1, 3).

[67] Bohr notes, for example, that "the very word 'experiment' refers to a situation where we can tell others what we have done and what we have learned" (PWNB, 2:72; cf. 39). Aage Petersen also recalls Bohr's following remarks: "There is no quantum world. There is only an abstract quantum physical description. It is wrong to think that the task of physics is to find out how nature is. Physics concerns what we can say about nature" (Petersen 1963, 305).

What Bohr calls common language is not identical to our "daily language," which, as he admits, is not always fully unambiguous. For this reason, he refers to the "special role played by mathematics" in physical science. According to Bohr, we can use mathematics not as "a separate branch of knowledge," but rather as "a refinement of general language" in order "to represent relations for which ordinary verbal expression is imprecise or cumbersome" (PWNB, 2:68; cf. 3:9). In particular, "just by avoiding the reference to the conscious subject which infiltrates daily language, the use of mathematical symbols secures the unambiguity of definition required for objective description" (PWNB, 2:68; cf. 70, 87).[68] In Bohr's view, the language of classical physics is a specific part of common language that has thus been made unambiguous by virtue of the use of mathematical symbols. The unambiguity of description and communication may therefore be achieved by means of "plain language, suitably refined by the usual physical terminology," that is, common language supplemented by the terminology of classical physics (PWNB, 3:3; cf. 11, 2:72).

Bohr conceives this linguistic requirement for unambiguity as correlative with the epistemological demand of the subject/object distinction. He stresses that, as in every other "field of experience," so also in quantum theory, we must draw a sharp line "between subject and object" for the purpose of unambiguous communication (PWNB, 2:91). This implies not only "a fundamental distinction between the measuring apparatus and the objects under investigation" (PWNB, 3:3),[69] but, more generally, "a sharp distinction between the observer and the content of the observations" (PWNB, 2:91). Moreover, under certain circumstances, we must make "different placings of such a separation" to avoid the ambiguity of the communication (PWNB, 2:92). Specifically, complementary phenomena such as those conditioned by space-time coordination and those conditioned by dynamical conservation must be treated with different placings of the subject/object distinction, but – precisely for this reason – can both be described unambiguously.

Further, Bohr goes on to discuss more specifically how to achieve the unambiguity of quantum-theoretical description. According to Bohr, "even when the phenomena transcend the scope of classical physical theories, the account of the experimental arrangement and the recording of observations must be given in plain language, suitably supplemented by technical physical terminology," and only in this way can the unambiguity of communication be ensured (PWNB, 2:72; cf. 39, 89). First, as regards the experimental arrangement, this requirement is fulfilled "by the use, as measuring instruments, of rigid bodies sufficiently heavy" to "permit the quantum to be neglected in their description," and thus to "allow a completely classical account of their relative positions and velocities" (PWNB, 3:3, 2:89; cf. 98). Second, "the recording of observations" under such experimental conditions – that is, information on 'phenomena' – must be "based on registrations obtained by means

[68] For comparison with Bohr's 'early' view of mathematics, see PWNB, 1:97.

[69] For parallel passages, see PWNB, 2:50, 55f., 80, 3:6, and 78. See also Faye (1991, 189).

of suitable amplification devices with irreversible functioning such as, for example, permanent marks on a photographic plate" (PWNB, 2:73; cf. 64).[70] In Bohr's view, this "closed character" of phenomena makes possible "all unambiguous information concerning atomic objects" (PWNB, 2:98, 3:3). This may also be seen as a condition for the usage of the term 'phenomenon' itself: One must reserve this word "solely for reference to unambiguously communicable information" (PWNB, 3:6).[71] A third and most pivotal point concerns the *relation* between the experimental arrangement and the phenomenon. By recourse to his redefinition of the phenomenon introduced in the 'middle' period, Bohr claims that unambiguous description may be achieved only "by including in the account of the phenomena explicit reference to the experimental conditions" (PWNB, 3:12). In this way, one can avoid "any appeal to the observing subject, which would hinder unambiguous communication of experience." This is analogous to the circumstance regarding relativity theory: "[T]he dependence of the phenomena on the reference frame of the observer" in no way renders the description ambiguous, but, on the contrary, including this dependence in the account of the phenomena secures the unambiguity of the description (PWNB, 3:7; cf. 10, 2:70).[72] Similarly, in quantum theory, "all subjectivity is avoided" precisely by including in the account of the phenomena the experimental condition set by the observer (PWNB, 3:7; cf. 3, 12). This being the case, "there is, strictly speaking, no new observational problem in atomic physics" (PWNB, 2:89; cf. 64, 3:3).

Bohr emphasizes that the complementary description of atomic phenomena fulfills the above series of requirements for unambiguity and objectivity. Although "one has sometimes seen in the notion of complementarity a reference to the subjective observer, incompatible with the objectivity of scientific description" (PWNB, 2:90f.; cf. NBCW, 7:420), he holds this misguided impression to be finally cleared up.[73] Far from having subjectivist implications, the complementarity description of phenomena is of "a perfectly objective character, in the sense that no explicit

---

[70] This focus on the permanency of the results of observation is not, however, in itself new to Bohr's thought. See, for example, letter from Bohr to Dirac, March 24, 1928 (NBCW, 10:495–96, on 495; cf. xxxvi).

[71] In this connection, it is also noteworthy that in the 'late' period Bohr tends to avoid using the expression 'uncontrollable interaction.' See Loren R. Graham, "The Soviet Reaction to Bohr's Quantum Mechanics," in Feshbach, Matsui, and Oleson (1988, 305–17, on 313). While still speaking of the "transfer, uncontrollable in principle, of momentum and energy" (PWNB, 3:11), Bohr may consider it misleading to call the interaction itself 'uncontrollable.' For, in his view, the way in which the phenomenon, including the object–instrument interaction, is conditioned by the experimental arrangement can itself be determined unambiguously, and is thus not uncontrollable.

[72] Bohr's comparison between relativity and quantum theory underwent considerable historical changes and fluctuations. During the 'early' and 'middle' periods, he primarily stressed "the new epistemological situation" brought about by quantum theory in *contrast* to relativity theory (PWNB, 4:85; cf. 2:19, 25, 41, 4:87; NBCW, 10:60). When at times discussing their commonalities, he focused on the innovative or revolutionary character marking both theories (see PWNB, 2:64, 4:85f.; NBCW, 7:317). In the 'late' period, however, he came to emphasize that neither of the two theories deviates from the classical ideal of objectivity and unambiguity.

[73] As we have seen, Bohr himself had earlier associated complementarity with the "subjective character of all experience" as well as with an unavoidable "ambiguity in our use of language"

reference is made to any individual observer and that therefore [...] no ambiguity is involved in the communication of information" (PWNB, 3:3). As he puts it in an interview conducted shortly before his death, complementary description is an "objective description" – indeed even "the only possible objective description" – of nature.[74]

This objectivist tendency of Bohr's thought finds its most prominent expression in the following passage in his 1954 lecture "Unity of Knowledge":

> The notion of complementarity does in no way involve a departure from our position as *detached* observers of nature, but must be regarded as the logical expression of our situation as regards objective description in this field of experience (PWNB, 2:74, my emphasis).

This remark seems to be so divergent from Bohr's earlier accounts of complementarity that one may question the coherence of his philosophical position over time. In fact, Bohr's close fellow physicist Wolfgang Pauli, who had basically accepted his complementarity argument, reacted critically to the above passage by saying that "no justification seems to me left for the way you use the phrase 'position of the detached observer,'"[75] because the quantum-mechanical description of nature implies "an abandonment of the idea of the isolation (detachment) of the observer from the course of physical events outside himself."[76] Bohr defended himself by responding that his phrase 'detached observer' had "a very definite meaning," namely, the "demand that the separation between the observing subject and the objective content of communication is clearly defined and agreed upon."[77] Although Bohr thereafter did not use again the phrase "detached observers,"[78] this can hardly be taken as a sign of a substantive change in his thought. Despite its being conspicuous in the mode of expression, the above passage does accord with many other remarks on complementarity that he made from around 1950 onward.

What we have seen so far about Bohr's 'late'-period idea of complementarity basically applies to fields other than quantum physics as well. In his account, "biological and psychological phenomena" may also be "comprehended within

(PWNB, 1:1, 19). While this does not mean that he had subscribed to subjectivism, it is worth reconsidering the question of how and in what sense the impression in question is misguided.

[74] Interview with Bohr conducted by Thomas S. Kuhn and Aage Petersen, November 17, 1962, in *Archive for the History of Quantum Physics*, p. 4.

[75] Letter from Pauli to Bohr, between April 6 and 26, 1954, in *Niels Bohr Archives: Bohr Scientific Correspondence*.

[76] Letter from Pauli to Bohr, February 15, 1955, in *Niels Bohr Archives: Bohr Scientific Correspondence*.

[77] Bohr continues as follows: "Just as Einstein himself has shown how in relativity theory 'the ideal of the detached observer' may be retained by emphasizing that coincidences of events are common to all observers, we have in quantum physics attained the same goal by recognizing that we are always speaking of well defined observations obtained under specified experimental conditions." Letter from Bohr to Pauli, March 2, 1955, in *Niels Bohr Archives: Bohr Scientific Correspondence*. See also Folse (1985, 212–14).

[78] In subsequent works, he speaks, for example, of "a complementary relationship which is connected with our position as observers of nature" (PWNB, 2:92; cf. 94, 99), avoiding the phrase 'detached observers.'

the framework of objective knowledge," marked by "the separation between the observer and the content of the communications" (PWNB, 2:91; cf. 101). Admittedly, unlike the case of quantum theory, "in a living organism, [. . .] a distinction between the measuring instruments and the objects under investigation can hardly be fully carried through" (PWNB, 2:92). Yet, stresses Bohr, "communication of biological experiences contains no more reference to the subjective observer than does the description of physical evidence" (PWNB, 2:101), so that "in biology just as in physics, we retain our position as *detached* observers" (PWNB, 2:76, my emphasis).[79]

Bohr's attitude toward psychological problems is somewhat more nuanced, yet not essentially different. He no longer speaks, as he had still done in the 1930s, of "the impossibility of distinguishing, in introspection, sharply between subject and object" (PWNB, 2:22).[80] Rather, similarly to the cases of other fields, "[e]very unambiguous communication about the state and activity of our mind implies [. . .] a separation between the content of our consciousness and the background loosely referred to as 'ourselves.'" Further, depending on the situation, one must and is able to make "a different placing of the section between subject and object" (PWNB, 3:12f.).[81] Specifically, to retain the subject/object distinction, "in every communication containing a reference to ourselves we, so-to-speak, introduce a new subject which does not appear as part of the content of the communication" (PWNB, 2:101). In this way, it is possible to "uphold the requirements of objective description to a great extent in human psychology" (PWNB, 2:76).

This view – and, by extension, Bohr's overall idea of complementarity in the "late" period – may be illustrated by the way in which he refers, in a 1960 speech, to Møller's novel *The Adventures of a Danish Student*. Immediately after citing the licentiate's remark on his "infinite retrogressive sequences of 'I's,'" which we have seen earlier, he quotes a response of his cousin, known as a "philistine":

> I cannot in any way help you in sorting your many "I"s. [. . .] My line is to stick to palpable things and walk along the broad highway of common sense; therefore my "I"s never get tangled up (cited in PWNB, 3:13).

This quotation is followed by Bohr's own comment that "[f]ortunately, the risk of falling into the deplorable situation of the licentiate is small in normal life, where we become gradually accustomed to coping with practical necessities and learn to communicate in common language what we need and what is on our mind" (PWNB, 3:13f.). This suggests that his thought in the 'late' period is no longer oriented

---

[79] It should be noted that, toward the end of his life, apparently in response to the recent development of molecular biology (as represented by the discovery of DNA), Bohr became rather reticent about the idea of complementarity in biology. Specifically, deviating from his long-held view on the physical analysis of the phenomena of life, he remarked that "we have no reason to expect any inherent limitation of the application of elementary physical and chemical concepts to the analysis of biological phenomena" (PWNB, 4:184; cf. 188). See Pais (1991, 443f.) and Beller (1999, 271).

[80] The quotation is from Bohr's 1937 lecture entitled "Biology and Atomic Physics" (PWNB, 2:13–22).

[81] For parallel passages, see PWNB, 2:52, 77, 81, 93, and 4:177.

toward the licentiate's interminably entangled 'I's,' which leads to a "bottomless abyss," but rather toward the overcoming of this state of affairs by going along the philistine's "broad highway of common sense" (see Folse 1985, 54).

In this chapter, I have outlined Bohr's idea of complementarity, tracing its development from 1927 until the end of his life. To be sure, just as any other account of the same or a similar subject, my description so far cannot claim to be purely neutral, but is already guided by some interpretive point of view of my own. Nevertheless I have not yet thematically presented my interpretation of the subject. Specifically, I have indeed at times suggested, but not inquired into, possible historical changes in his philosophical position. It is not until Chapter 4 that I will set forth an analysis of Bohr's complementarity in both its conceptual structure and its historical development. Prior to that, I wish to devote the next chapter to a review of various interpretations of complementarity by a number of past as well as contemporary commentators.

Chapter 3
Prior Interpretations of Complementarity

Bohr's idea of complementarity is generally characterized as a key component of what is called the Copenhagen interpretation of quantum theory. As quantum mechanics was firmly established as a cornerstone of modern physics, and as the Copenhagen interpretation became predominant over the competing accounts, Bohr's complementarity became widely recognized as an essential constituent of the 'orthodoxy.' This does not mean, however, that his point of view was universally accepted or always clearly or unambiguously understood.[1] Rather, there was and still is the widespread impression that his complementarity argument is "obscure" and refractory to clear-cut analysis – an impression which has at times led to the suspicion of 'mysticism' as mentioned earlier (Folse 1985, 29; cf. 108).[2] This seems to be one of the main reasons why Bohr's complementarity has been subject to a noticeably broad variety of interpretations. A number of physicists, philosophers, and historians of science have engaged in the reading of Bohr, and presented accounts that greatly differ from, or conflict with, each other with regard to the meaning and implications of complementarity and its historical path of development.

In this chapter, I will survey prior interpretations of Bohr's complementarity by a wide range of commentators. In Section 3.1, I review some early interpretive accounts by physicists, especially within his Copenhagen-oriented circle. Section 3.2 is devoted to a survey of various and often conflicting interpretations of Bohr's philosophical position that largely revolve around the axis of realism and anti-realism. This is closely linked to the subject of Section 3.3: the debate on the historical development of his philosophical thought. Finally, in Section 3.4, I address some attempts to situate Bohr's complementarity in the historical context of modern philosophy, notably in its relation to Kantian philosophy.

---

[1] In his last interview, Bohr remarked with some frustration that "[n]o man who is called a philosopher really understands what one means by complementarity description." *Archive for the History of Quantum Physics*, p. 4. According to Ulrich Röseberg, "[m]any physicists and nearly all philosophers of his time disliked Bohr's complementarity argument" (1995, 117).

[2] See also Held (1994, 871). While favorable commentators speak of the "subtlety" of Bohr's thinking, the critic Mara Beller, for instance, characterizes his philosophy as "obscure and inconsistent" (1999, 271, 275).

M. Katsumori, *Niels Bohr's Complementarity*, Boston Studies in the Philosophy of Science 286, DOI 10.1007/978-94-007-1748-0_3,

3.1 Early Interpretations by Physicists

As we saw in the previous chapter, Bohr's idea of complementarity in quantum theory primarily refers to the relation between space-time coordination and the claim of causality (or dynamical conservation), while he only rarely and derivatively spoke of the complementarity of particle and wave or of position and momentum. However, these latter notions of complementarity – particle-wave as well as position-momentum complementarity – became prevalent among physicists of his times. In 1933, Wolfgang Pauli gave the following definition of the term 'complementary': "If [. . .] the possibility of the use of *one* classical concept stands in an exclusive relation to that of *another*, with Bohr we call these two concepts (e.g. position and momentum coordinates of a particle) *complementary*."[3] In a later writing, he mentioned both "momentum and position" and "wave and particle" as major examples of complementary classical concepts.[4] This line of interpreting the concept of complementarity was widely shared by physicists outside the Copenhagen circle as well. David Bohm, for example, the chief advocate of the hidden-variable theory as against the Copenhagen interpretation, describes Bohr's complementarity as follows: "the principle of complementarity states that we are restricted to complementary pairs of inherently imprecisely defined concepts, such as position and momentum, wave and particle, etc." (1957, 93).

Yet some of Bohr's closest collaborators and followers, specifically Heisenberg and von Weizsäcker, did correctly recognize that his concept of complementarity centered on the relation between space-time coordination and the claim of causality. This does *not* mean, however, that by this relation between space-time and causality they understood exactly the same as Bohr did. Rather, as it will turn out, in their apparently accurate accounts of Bohr's complementarity, their points of view tacitly diverged from the latter. In what follows, I will take up in turn Heisenberg's and von Weizsäcker's accounts of complementarity and suggest how they deviate from Bohr's position.

In his 1930 book *Physikalische Prinzipien der Quantentheorie* [*The Physical Principles of the Quantum Theory*], Werner Heisenberg devoted a section to an exposition of the idea of complementarity.[5] His account, one of the earliest major interpretations of complementarity, might appear to be, and in part actually *is*, a faithful reproduction of Bohr's thought (see Cassidy 1992, 265). As Heisenberg puts

---

[3] Wolfgang Pauli, *Handbuch der Physik*, Vol. XXIV/1 (Berlin: Springer, 1933), p. 89, cited in von Weizsäcker (1976, 284).

[4] Pauli (1961, 28). This is not to say, however, that Pauli's conception of complementarity is *restricted* to these notions. For his discussion of complementarity, see "Die philosophische Bedeutung der Idee der Komplementarität" (1961, 10–17).

[5] In Chapters 1 and 2, I touched on some of the differences between Bohr's and Heisenberg's physical-scientific as well as philosophical views. Although it would be worth further examining the commonalities and differences between the two physicists' thought, here I restrict myself to the question of how Heisenberg interpreted Bohr's idea of complementarity. For recent accounts of Heisenberg's scientific and philosophical thought, see Schiemann (2008) and Camilleri (2009).

it, the principle of causality rests on "the assumption that it is possible to observe the phenomena without appreciably influencing them" (1930, 48/62f.). In atomic physics, however, any observation involves "a finite, to a certain degree uncontrollable, disturbance (*Störung*)" (1930, 48/63, trans. mod.).[6] Since "any space-time description of a physical process is conditioned by the observation of the process," it follows that "the space-time description of processes on the one hand and the classical law of causality on the other represent complementary, mutually exclusive features of the physical process" (1930, 48/63f., trans. mod.). Up to this point, there seems to be no conceptual divergence from Bohr's 'early' formulation of complementarity.

Heisenberg goes on to say, however:

> To this situation [of complementarity] there corresponds in the formalism of the theory the circumstance that there indeed exists a mathematical schema of quantum theory, but that this schema cannot be interpreted as a simple connection of things in space and time (Heisenberg 1930, 48/64, trans. mod.).

Heisenberg then refers to this circumstance as the "complementarity of space-time description on the one hand and causal connection on the other" (1930, 48/64, trans. mod.). Does this also accord with Bohr's view? As we saw in the previous chapter, Bohr conceives the claim of causality (associated with state definition) as equivalent to the use of the dynamical conservation laws, that is, the laws of conservation of energy and momentum. In contrast – as pointed out by Shingo Fujita (1991, 106f.) and later by Kristian Camilleri (2009, 113ff.) – what Heisenberg now means by "causal connection" is *not* dynamical conservation, but the "mathematical schema of quantum theory," and he thus reconceptualizes complementarity as the relation between a space-time description and a description in terms of that mathematical schema. As indicated in Section 2.1 (note 8), this reading may have been suggested by Bohr's phrase "the definition of the state of a physical system" (PWNB, 1:56), which, to Heisenberg, seemed to refer to the quantum-mechanical state as expressed by the state or wave function. Whether we should regard this construal by Heisenberg as a sheer "misinterpretation" (Fujita 1991, 107) or rather as a kind of productive conceptual expansion, there is no doubt that he imparted to the concept of complementarity a new meaning that had not been intended by Bohr.

From then onward, as it will turn out, the above displaced meaning of complementarity came to constitute Heisenberg's major line of understanding of the subject. Let us take a brief look at his discussion of Bohr's complementarity in his later work *Physics and Philosophy: The Revolution in Modern Science*, published in 1958. After touching on the complementarity of particle and wave pictures and the complementarity of "the knowledge of the position of a particle" and "the knowledge of its velocity or momentum," he proceeds to the relation between space-time and causality, stating that "[t]he space-time description of the atomic events is

[6] Since the existing translation of the *Physikalische Prinzipien der Quantentheorie* by Eckart and Hoyt greatly deviates from the original, I have considerably changed the translation of this and the following passages or rather partly translated them anew myself.

complementary to their deterministic description." By "deterministic description" here he means the following: The way in which the state function or "probability function" changes in time "is completely determined by the quantum mechanical equation," which does not allow a description in space and time. On the other hand, observation "enforces the description in space and time but breaks the determined continuity of the probability function by changing our knowledge of the system" (Heisenberg 1958, 49f.). We can thus see that while in his earlier account he had still referred to the "*classical* law of causality" (my emphasis) as well, Heisenberg has now fully converted Bohr's complementarity of space-time and causality into the relation between a space-time description and a description in terms of the probability function. This move appears to be correlative with Heisenberg's mature ontology revolving around the notion of "potentiality," which accords a kind of 'reality' to the mathematical formalism of quantum mechanics.[7]

With this interpretation by Heisenberg in mind, let us now proceed to a closely related but philosophically more inventive account of complementarity, offered by his onetime student and associate Carl Friedrich von Weizsäcker. In his 1955 paper "Komplementarität und Logik [Complementarity and Logic]" (included in 1976, 281–331), von Weizsäcker introduced a distinction between what he terms "parallel" and "circular" complementarity. By "parallel complementarity" he means the complementary relation between different descriptive concepts "of the same level" – such as position and momentum or wave and particle. This notion of complementarity, of which Bohr at times speaks, has been commonly and widely accepted among physicists (1976, 284ff.). Von Weizsäcker argues, however, that this parallel complementarity is not the only conception of complementarity, and not the one primarily held by Bohr.

Like Heisenberg, von Weizsäcker considers Bohr's original idea of complementarity to be focused on the relation "between space-time description and the claim of causality," and, further, identifies this relation with the relation "between the description of nature in terms of classical concepts and in terms of the ψ function" (1976, 293). He links this view, however, to the following argument of his own. In von Weizsäcker's account, the complementarity of space-time description and the claim of causality (as thus interpreted) is not a parallel complementarity, because the complementary relata in this case do not stand on the same conceptual level. On the one hand, the quantum-mechanical description of nature in terms of the wave function presupposes the classical space-time concepts in the sense that it may be meaningfully connected with experience only by means of the latter. On the other hand, these classical concepts are in turn subject to modification by quantum-mechanical findings. In more general terms, von Weizsäcker continues:

[7] Elsewhere in the same book, Heisenberg remarks: "The 'thing-in-itself' is for the atomic physicist, if he uses this concept at all, finally a mathematical structure; but this structure is – contrary to Kant – indirectly deduced from experience" (1958, 91). This ontology diverges from Bohr's thought as well as from Heisenberg's own earlier positivist tendency. Needless to say, this by no means implies that Bohr's position is akin to positivism.

> Perhaps we may designate as the "circle of knowledge" the circumstance that our knowledge has a priori presuppositions which nevertheless become corrected and reinterpreted retroactively by substantive findings obtained with their own help. Since Bohr's conception we have just discussed starts from the recognition of this circle, let us term the complementarity of which he speaks here "circular complementarity" (von Weizsäcker 1976, 294).[8]

In von Weizsäcker's view, this is precisely "Bohr's original conception" of complementarity (1976, 290). Furthermore, this circular complementarity holds in many analogous instances outside quantum physics discussed by Bohr. As we have seen earlier, Bohr points, for example, to the "complementarity between the definition (or analysis) of a concept and its immediate use." According to von Weizsäcker, this complementarity is also circular, because "I understand concepts better if I analyze them by a definition; but I can define only if I already immediately understand concepts" (1976, 294).

This account by von Weizsäcker of parallel/circular complementarity was not, however, accepted by Bohr himself. In a postscript to the same paper, von Weizsäcker notes that, in discussing Bohr's complementarity of space-time description and the claim of causality, he had drawn on Heisenberg's interpretation, according to which the claim of causality refers to "Schrödinger's wave theory in configuration space" (von Weizsäcker 1976, 330; cf. 293). Bohr, he continues, "objected to this interpretation by letter," contending that "complementarity can hold only between phenomena," that is, "between concepts designating something perceptible in experiments," while "the Schrödinger wave function is merely an algorithmic quantity (*Rechengröße*) and signifies no phenomenon." This objection by Bohr rests on a notion of parallel complementarity, and, with this notion, "the complementarity between space-time description and causality reduces to the well-known complementarity between position and momentum or time and energy." Von Weizsäcker thus admits that he "misunderstood Bohr" in this regard. This does not mean, however, that he abandons the concept of circular complementarity itself.[9] To be sure, he continues, "one could restrict the concept of complementarity to the cases in which I spoke of parallel complementarity, and, in other cases, could speak of a circular relation without using the term complementarity" (1976, 330). Yet, considering the fact that "almost all examples of complementarity outside physics given by Bohr" are of a circular nature, the designation of circular complementarity might still be appropriate in these cases (1976, 331).

So far we have reviewed some early interpretations of Bohr's complementarity by contemporaneous physicists, mainly by Heisenberg and von Weizsäcker, and

---

[8] See Jammer (1974, 103), Folse (1985, 269f.), Honner (1987, 58f.), and Held (1994, 883f.). This notion of the "circle of knowledge" is not restricted to von Weizsäcker's *interpretation* of Bohr's complementarity, but constitutes one of the major concepts of his own philosophical thought. See von Weizsäcker (1971, 82/62, 105/83).

[9] As Jammer comments, von Weizsäcker was not satisfied by Bohr's response, because, following Heisenberg rather than Bohr, he regarded the formalism of quantum mechanics as much more than a mere algorithm (Jammer 1974, 104).

suggested their conceptual divergences from Bohr's position. I now wish to conclude this section with brief comments from my own interpretive point of view. To be sure, in no period of his career did Bohr mean by "the claim of causality" description in terms of the wave function, or embrace the notion of the complementarity of classical space-time description and the quantum-mechanical formalism. To this extent, Heisenberg's and von Weizsäcker's interpretations of complementarity did contain, as the latter admitted, a misunderstanding. From this it does *not* follow, however, that Bohr's complementarity was always restricted to 'parallel complementarity.' We should take into account the fact that the above exchange of views between von Weizsäcker and Bohr took place in the 1950s, when Bohr was in the final stage of his intellectual career. In Chapter 4, I will show how Bohr's complementarity in earlier periods contains a conception that somewhat resembles von Weizsäcker's circular complementarity without quite according with it. It should also be noted that von Weizsäcker's concept of circular complementarity is reminiscent of what is known as the hermeneutic circle, a key concept of modern hermeneutic philosophy. In Chapter 5, I will investigate precisely the conceptual intersections between Bohr's complementarity and hermeneutic philosophy. My analysis will suggest, however, that while Bohr's thought is in some important respects parallel to hermeneutic philosophy, their similarities do not center on the concept of the circle in von Weizsäcker's sense.

The interpretations of complementarity reviewed in this section may in a sense be characterized as products of an era in which many physicists still retained keen interest in Bohr's philosophical thought and, more generally, in the philosophical foundations of quantum theory. Gradually, however, as the majority of physicists became content to *use* quantum mechanics as a firmly established theory, the philosophical import of complementarity was driven to the periphery of their intellectual interest, and the concept of complementarity itself became treated even like an almost superfluous appendage to the exposition of quantum theory (see Jammer 1974, 247f.). This tendency appears to have continued up until now. This is not to say, however, that people's interest in Bohr's philosophical thought has generally faded away. Rather, his thought has increasingly drawn attention among the historians and philosophers of science, becoming – along with Einstein's work in the same century – a major subject of research in their fields. In the next three sections, I will accordingly survey some of the most important of these philosophical and historical inquiries into Bohr's complementarity.

3.2 Complementarity and the Realism Debate

During his lifetime and beyond, Bohr's idea of complementarity was long and most commonly viewed – whether favorably or critically – as opposed to realism (see von Weizsäcker 1971, 407/328f.). A number of early commentators, notably Karl Popper and Mario Bunge, considered Bohr's position to be subjectivist and positivist anti-realism (see Honner 1987, 14), although their concern was not so much with Bohr's thought as such, but rather with the so-called Copenhagen interpretation in general.[10] In his article "Quantum Mechanics without 'the Observer,'"[11] Popper argues that the Copenhagen interpretation is characterized by the special significance it attaches to the "observer" or the "subject," based on the claim that "atomic theory takes its peculiar character largely from *the interference of the subject or the observer (and his 'measuring agencies') with the physical object under investigation*" (1982, 40; cf. 6). This view is exemplified not least by Bohr's remark in his response to EPR that "the *finite interaction between object and measuring agencies* [...] entails the necessity of a final renunciation of the classical ideal of causality and a radical revision of our attitude towards the problem of physical reality" (PWNB, 4:75).[12] In Popper's account, this renders quantum mechanics a representation not of particles, but of "our knowledge, our observations, or our consciousness, of particles" (1982, 35). Thus responsible for "the intrusion of the observer, or the subject, into quantum theory" (1982, 50),[13] the Copenhagen interpretation and, in particular, Bohr's complementarity as a key component thereof, may be characterized as a "subjectivistic and anti-realistic interpretation of quantum theory" (1982, 85; cf. 5, 64).[14]

Unlike this kind of rather simplistic account, there have recently been more sophisticated approaches to Bohr's thought with due attention to the philosophical differences between Bohr, Heisenberg, and other Copenhagen-associated physicists. On one side, there are commentators who, in contrast to Popper, reinterpret Bohr's position as oriented to realism (see Faye and Folse 1998, 13). Henry Folse, in particular, maintains that Bohr's complementarity does not imply phenomenalism, but rather realism with a belief in the independent existence of atomic objects.[15] In

[10] Mario Bunge interpreted and critically assessed Bohr's thought as antithetical to realism (see Folse 1985, 262). He characterizes Bohr's complementarity and, by extension, the "Copenhagen doctrine" in general by various designations such as idealism, positivism, phenomenalism, and "semi-subjectivism" (Bunge 1955, esp. 7; 1988, 403). See also his "The Turn of the Tide," in Bunge (1967, esp. 4).

[11] This writing, which serves as the "Introduction" to Popper (1982, 35–95), is a revised and expanded version of an article with the same title included in Bunge (1955, 7–45).

[12] Popper also cites Heisenberg's following remark: "The conception of objective reality [...] has thus evaporated [...] into the transparent clarity of a mathematics that represents no longer the behaviour of particles but rather our knowledge of this behaviour" (Heisenberg 1971, 101; see Popper 1982, 85).

[13] From his own philosophical viewpoint, which is definitely realist, Popper attempts, as it were, "to exorcize the ghost called 'consciousness' or 'the observer' from quantum mechanics" (1982, 35). For this purpose, he seeks to refute "the dualism of particle and wave" and what he calls "the subjective interpretation of probability," on both of which, in his view, the Copenhagen interpretation rests (1982, 85).

[14] It is worth noting that, in the 1930s, logical positivists such as Philipp Frank and Otto Neurath characterized Bohr's complementarity neither as realist nor as subjectivist, but as close to their own positivist position (see Beller and Fine 1994, 19; Faye and Folse 1998, 7ff.). See also Section 3.4 of this book. The designation of Bohr's thought or at least part of it as "subjectivist" has been repeated, however, by some commentators up until now. Toward the end of Chapter 6, I will touch on this issue with reference to the Science Wars.

[15] As Jan Faye comments, "Henry Folse was one of the first philosophers to maintain that Bohr's philosophy is to be understood realistically in spite of its anti-realistic flavour" (Faye 1991, 203).

his most comprehensive work on the subject, *The Philosophy of Niels Bohr: The Framework of Complementarity* (1985), Folse presents the following set of arguments for his realist reading of Bohr. First, he refers to Bohr's view that the quantum postulate is the result of an "empirical discovery" that "we are forced to adopt [...] by the way the world is." This stands in sharp conflict with the phenomenalist position. For the phenomenalist would maintain that since "the quantum postulate refers to atomic systems," which are not phenomenal objects, it is merely "an intellectual construction for describing certain phenomena." From this it would follow that there may be another theoretical construction which allows a description of the same phenomena without appeal to discontinuous actions. Folse stresses, however, that "this is most definitely a view Bohr denied." We can thus see that "Bohr's strong belief that such an alternative was impossible [...] belies the phenomenalist interpretation of [his] intentions" (1985, 234).[16]

Folse's second argument runs as follows: On the phenomenalist assumption, the atomic system "must be considered a construct of the theory," whereas "[t]he observing system is given as a phenomenal object." This would entail that "observation is an interaction between a theoretical construct and a phenomenal object." Such a view, however, "seems patently absurd,"[17] and in fact is not held by Bohr (1985, 232; cf. 243–45; Faye 1991, 205). Moreover, while, for the phenomenalist, the observing system and the atomic system "must belong to distinct ontological orders," Bohr's complementarity holds that the distinction between the two is an arbitrary one (Folse 1985, 232).

In his "third line of reasoning," Folse starts by pointing out that, from the late 1930s on, when speaking of complementary phenomena or complementary kinds of information, Bohr "clearly indicates that they are 'complementary' because they are both about 'the same object'" (Folse 1985, 237). As quoted in Section 2.3 of this book, Bohr says, for instance: "Information regarding the behaviour of an atomic object obtained under definite experimental conditions may [...] be adequately characterized as *complementary* to any information about *the same object* obtained by some other experimental arrangement excluding the fulfilment of the first conditions" (PWNB, 2:26, the latter emphasis is Folse's; see Folse 1985, 237f.). According to Folse, this "same object" cannot be the phenomenal object, because the latter cannot be said to exhibit different appearances under different experimental arrangements. Therefore:

> The only alternative is to assume that Bohr intends "the atomic object" which is "the same" throughout both experimental interactions to be the object which causes these complementary phenomena (Folse 1985, 238).

In other words, "although Bohr avoids talk of the nature of an independent reality, his talk of different phenomena as complementary presupposes the existence of the

---

[16] As a support of his view of Bohr's position as realist, Folse also calls attention to his remark that "every doubt regarding the reality of atoms has been removed" (PWNB, 1:103; cf. 93, 2:24; NBCW, 10:55).

[17] Here Folse does not seem to give an adequate reason for rejecting the possible phenomenalist view according to which the observational interaction is itself a theoretical construct.

atomic system as an independent reality" (1985, 239; cf. 151).[18] Based on the series of arguments outlined so far, Folse concludes – in diametrical opposition to Popper's account – that Bohr's philosophical orientation should be characterized as realist.

Some other researchers, however, seek to avoid choosing simply between the poles of realism and subjectivist anti-realism as the interpretation of Bohr's position. Jan Faye, in particular, in his 1991 book *Niels Bohr: His Heritage and Legacy – An Anti-Realist View of Quantum Mechanics*, offers an account of Bohr's thought neither as realism nor as phenomenalism or positivism, but as what he calls "objective anti-realism." Faye starts by criticizing the realist reading of Bohr as represented by Folse's (Faye 1991, 203–11; cf. 195). As Faye admits, Folse is "quite right in saying that Bohr's notion of complementarity is incompatible with phenomenalism." Yet this does not mean that Bohr was "a realist believing in the existence of objects which belong to the transphenomenal, noumenal sphere" (1991, 206; cf. 204). Specifically, while remarking that "every doubt regarding the reality of atoms has been removed," "Bohr does not say anything about the nature of the reality belonging to the atom" (1991, 209). Folse's claim that Bohr regards the atomic system as "an isolated, independent object having some inherent states" is, according to Faye, untenable (1991, 205). Bohr denies such a realist notion, and therefore also the idea that "the aim of quantum mechanics (or of any other physical theory) is to explain the phenomena in terms of an underlying, hypothetical reality" (1991, 217). This is illustrated by his remark in 1953 that "it is difficult for me to associate any meaning with the question of what is behind the phenomena."[19] On the other hand, continues Faye, it is also Bohr's view that "the physical world is objective in the sense that the content of experience, insofar as what we observe may be given a causal space-time description, exists independently of the mind" (1991, 206). This suggests that "the view of complementarity [...] is different from both realism and idealism," and that it holds "some intermediate position" (1991, 217).

Faye develops this line of interpretation with the aid of the following definitions of realism and anti-realism. Drawing broadly on Michael Dummett's formulation, he defines the central tenets of realism as follows: (1) "the world exists independently of our minds;" and (2) "truth is a non-epistemic notion," that is, a statement is true or false "independently of our power to establish which of these values it is." In contrast, continues Faye, anti-realism is the position according to which "a statement possesses a determinate truth-value only if this value can be established." Realism and anti-realism are thus opposed to each other primarily in terms of the "transcendence" versus "immanence" of truth conditions (1991, 198). Faye emphasizes, however, that anti-realism is in no way a unitary position, but may be

[18] In his later works, Folse partly modifies his realist reading of Bohr. Distinguishing several different principles of realism, he maintains that Bohr's "position is close to the realist forces," although it "cannot be located on either realist or anti-realist side." Henry Folse, "Bohr's Framework of Complementarity and the Realism Debate," in Faye and Folse (1994, 119–39, on 125). See also Folse (1996).

[19] Letter from Bohr to Born, March 10, 1953, in *Niels Bohr Archives: Bohr Scientific Correspondence*, p. 25 (cited in Faye 1991, 210).

subdivided into at least two distinct, partly opposing variants. One of them, which he designates as "subjective anti-realism," is the exact reverse of realism, namely the view that (1) "no mind-independent world exists" and that (2) "truth is dependent upon our cognitive faculty." This includes positions known as idealism and phenomenalism. The other variant of anti-realism, which he terms "objective anti-realism," shares with realism the view that (1) "there is an objective, mind-independent world," but holds – against realism – that (2) "truth is related to our cognitive powers" (1991, 199). In other words, objective anti-realism, just as realism, "operates with a notion of objectivity," and yet, unlike realism, maintains that only decidable statements, and not undecidable ones, possess "investigation-independent truth values" (1991, 200).

It is, Faye stresses, precisely this objective anti-realism that constitutes Bohr's philosophical position (1991, 217, 225–36).[20] On the one hand, in thinking of "atomic objects as real mind-independent entities," Bohr breaks with subjective anti-realism. On the other, he is not a realist, either, "because he denied that quantum theory could in principle provide us with any true assertions of [the] atomic object other than what can be obtained on the basis of direct observation" (1991, 231f.). This means, Faye concludes, that Bohr's position is nothing other than objective anti-realism as defined above.[21]

Let us briefly look at yet another interpretation, offered by Dugald Murdoch, which, in a manner somewhat similar to but different from Faye's, finds in Bohr's complementarity both realist and non-realist elements. In his 1987 book *Niels Bohr's Philosophy of Physics*, Murdoch maintains, on the one hand, that "Bohr's philosophy of physics" has a realist tendency, not in the sense of realism as defined by Dummett, but in the sense of "a weaker form of realism" (1987, 200, 213, 215f.). That is, Bohr not only posits independent physical reality, but also holds that "the truth-value of an experimental statement, when it has one, is independent of our knowledge and perception," although "it is not also independent of our ability to know or means of knowing its truth-value" (1987, 215).[22] On the other hand, Murdoch also points to a non-realist, instrumentalist component in Bohr's thought insofar as the latter "regards physical theories *primarily* as tools for the conceptual organization of our sensory experience." More specifically, Bohr's theory of meaning, according to which "the making of a measurement is a necessary precondition

[20] Faye traces Bohr's objective anti-realism back to the influence of his philosophical mentor Høffding, whose position Faye also characterizes as objective anti-realism (Faye 1991, 215). On commentators' accounts of Høffding's possible influence on Bohr, see Section 3.4.

[21] More recently, Faye recharacterizes Bohr's philosophical position as a conjunction of 'entity realism' and 'theory antirealism' (Faye 2002a). Elsewhere, drawing on the views of Ian Hacking and Nancy Cartwright, he also discusses entity realism and theory realism in a general context of the philosophy of science (2002b, 179ff.).

[22] Specifically, Murdoch maintains that, in Bohr's view, "successful observation or measurement reveals the objective, pre-existing value of an observable" (1987, 107). Faye rejects this interpretation by Murdoch, while assessing favorably the latter's overall reading of Bohr (see Faye 1991, 229f.).

of the meaningful applicability of a physical predicate," rests on the "pragmatist" view that "the ascription of an exact position and an exact momentum to an object at the same time is meaningless [...] because the ascription has no practical consequences whatever" (1987, 222, 224). In this way, pragmatism serves as the basis of the "instrumentalist component in Bohr's philosophy" (1987, 225). Murdoch thus concludes that "certain realist and non-realist components are subtly interwoven in [Bohr's] philosophy," and characterizes his position as "instrumentalistic realism" (1987, 200, 222).[23] We can see that, while, similarly to Faye, avoiding both extremes of strong realism and subjective anti-realism as the interpretation of Bohr's position, Murdoch opts for a *mixed* philosophical standpoint rather than an intermediate position as in Faye's account.[24]

In this section, reviewing a variety of interpretations of Bohr's complementarity by historians and philosophers of science, we have seen how they are divided on the question of what philosophical position – along the axis of realism and anti-realism – is the appropriate designation of his thought. Indeed, a large part of the debate among commentators in the last few decades has revolved around such interpretive differences and oppositions regarding the issue of realism and anti-realism.[25] So far, however, I have abstracted from one possible thematic dimension constitutive of such conflicting readings of Bohr: the question as to whether his philosophical orientation underwent diachronic changes. In fact, my discussion in this section has been largely restricted to interpretations that see Bohr's thought as more or less constant over time.[26] As it will turn out, however, there are also commentators who think otherwise in this regard, and for whom this historical dimension is essentially relevant to the realism/anti-realism issue as sketched above. I will accordingly devote the next section to a review of various approaches to the question of the historical development of Bohr's complementarity in close conjunction with the subject matter of the present section.

3.3 Possible Diachronic Changes in Bohr's Thought

One of the points at issue in the interpretation of Bohr's thought is the question of whether and, if so, how his idea of complementarity or his general philosophical outlook changed over time from his 1927 Como lecture until the end of his life. Many commentators view Bohr's complementarity as basically unchanged throughout his

---

[23] Murdoch notes that "the realist component and the instrumentalist component are, so to speak, complementary sides of the phenomenon that is Bohr" (1987, 222).

[24] Another commentator, John Honner, characterizes Bohr's "philosophy of science" as a kind of realism or "relative realism," while simultaneously pointing to a "transcendental" character of his approach (1987, 41, 151, 9). As will be mentioned in Chapter 6, Karen Barad also interprets Bohr's philosophical position as realism, but as a "nonrepresentationalist" kind of realism (2007, 56).

[25] A number of interpretations of Bohr's philosophical position that fall under this category are included in Faye and Folse (1994).

[26] Here some reservations are needed for the case of Faye, whose views, as we will see below, may be characterized as 'intermediate' regarding this historical question as well.

career, seeing in his varying presentations no more than a "sharpening of terminology" or "refinement of argumentation."[27] Others, however, recognize certain alterations in his philosophical thought, while differing from each other in what degree or kind of significance they attach to these alterations (see Faye and Folse 1998, 2).

Most commentators recognize at least certain terminological changes in Bohr's complementarity argument – particularly with regard to the notion of "disturbance." Distinguishing between two consecutive periods separated by his 1935 debate with EPR, many share the view that his pre-EPR use of expressions such as 'the disturbance of phenomena by observation' tended to suggest a philosophically misguided notion – misguided from the point of view of his own real intent. Folse, for example, renders this notion as follows: that "the object of observation really does have the properties that would define its classical mechanical state, but the observing interaction disturbs that state in a way which makes knowledge of them impossible" (1985, 155). In basic accord with Folse on this point,[28] Faye makes more manifest the implication of this misguided notion: that "behind the experiential manifestations of the atomic object there must exist an object with its own inherent states" (1991, 193). This kind of critical viewpoint on Bohr's pre-1935 terminology is shared by a wide range of commentators, including Mara Beller, for instance, whose overall interpretation of his thought, as we will see below, differs greatly from Folse's as well as Faye's.[29] In this way, many researchers with various interpretive orientations agree that Bohr's earlier disturbance language misleadingly suggests the (perhaps naïvely) realist notion that there exist atomic objects possessing inherent properties which are only *posteriorly* subject to disturbance. It is worth noting that this line of critique of the disturbance terminology is *not* the same as Bohr's own later critique thereof. As we saw in Section 2.3, his later critical argument is directed not so much against the notion of atomic objects with inherent properties, but rather against the notion (as suggested by the term 'disturbance') that the observer "influences the events which may appear under the conditions he has arranged" (PWNB, 2:51). Here, however, I refrain from entering further into this issue of Bohr's 'early' disturbance language, which I will examine in Chapter 4 from my own interpretive point of view.

Despite their above broad agreement on Bohr's pre-EPR terminology, the commentators' views diverge on the question of how to see his subsequent revisions of the complementarity argument, marked by the rejection of the notion of disturbance.

[27] Jørgen Kalckar, "Introduction," in NBCW, 7:249–87, on p. 259.

[28] Faye argues more specifically that "[i]n the Como paper and in the other early papers the phenomenon is what is being disturbed when it interacts with the observing instrument." Bohr at times even speaks "as if atomic phenomena possess properties which constitute classical states but which are disturbed by the measuring interaction in such a way that knowledge of the states of the atomic phenomenon becomes impossible" (Faye 1991, 193; cf. 135).

[29] According to Beller, the concept of disturbance, which Bohr adopted from Heisenberg, inconsistently "presupposes the existence of objective exact values that are changed by measurement" (Beller 1999, 200). Before 1935, this "idea of disturbance, of 'interference' with the course of phenomena, underlay Bohr's writings on both objectivity and causality" (1999, 158).

In Folse's account, although "there were very real changes in [Bohr's] manner of speaking," it is "misleading to represent these changes as a modification of Bohr's viewpoint," since the revised argument "was already implicit in the Como paper" (1985, 154).[30] For Folse, Bohr's argumentative revisions are no more than a "refinement" (1985, 142) within the overall framework of realism as we saw it in outline in the previous section. This realism is not a naïve classical realism as it might be suggested by his earlier disturbance language, but is to be understood as the position that the atomic system exists as an independent reality and yet is devoid of classically definable inherent properties.

Unlike Folse, Faye maintains that while, prior to the EPR challenge, Bohr's complementarity tended to have certain realist elements, his subsequent formulations moved away from this tendency. According to Faye, in and through his debate with EPR, Bohr became aware of the necessity to change his argumentation in order to avoid the above classical realist implication of the disturbance language (see Faye 1991, 193). Criticizing the tendency of "most philosophers" to read Bohr "as if the arguments for his philosophical position had remained the same from 1927 to his death" (1991, 169; cf. xix, xxf.), Faye contends that "little by little Bohr made certain fundamental revisions in his terminology as well as with respect to the underlying philosophical arguments of complementarity." While he did not change his philosophy dramatically, he altered "some of the basic arguments in support of it." More specifically, "[a]fter 1935 his grounds for asserting complementarity were not so much epistemological as they were conceptual or semantical" (1991, 186). That is, he came to derive the impossibility of simultaneously measuring two conjugate variables from the "indefinability thesis" – the claim of the impossibility of simultaneously *defining* them – while he previously proceeded the other way round by the use of the notion of disturbance (1991, 185; cf. 188).[31] In this way, Bohr eliminated the disturbance terminology and its "misleading" realist implications from his complementarity argument (1991, 205), thus rendering his philosophical position more coherent – a coherent "objective anti-realism" in the sense outlined in the previous section. In short, concludes Faye, Bohr underwent a philosophical development from an incoherent objective anti-realism mixed with realist elements toward a coherent objective anti-realism (see 1991, 193).

Further, however, there are researchers who regard Bohr's philosophical path not merely as steps toward a more coherent position, but as involving an essential turn or transformation. Such commentators, though not in the majority, seem to have been increasing in number and gaining influence in recent years.[32] According to Arthur Fine and Mara Beller, in particular, Bohr held realist-oriented notions

[30] From a different angle, Léon Rosenfeld also denies a post-EPR change in Bohr's philosophical position, saying that "[his] refutation of Einstein's criticism does not add any new element to the conception of complementarity" (1967, 129).

[31] As Faye comments, the 'indefinability thesis' is originally Dugald Murdoch's term (Faye 1991, 168). See Murdoch (1987, 139).

[32] Faye and Folse also appear to have come somewhat closer to this group of commentators. This may be seen particularly in their joint account (in 1998) of Bohr's relation to logical positivists during the 1930s, as will be described in Section 3.4.

before his confrontation with EPR, but through this debate turned to anti-realist *positivism*. In his 1986 book *The Shaky Game: Einstein, Realism, and the Quantum Theory*, Fine made the following points: In his response to EPR, "Bohr simply identifies the attribution of properties [of an atomic system] with the possible types of predictions of future behaviour," and this view is "virtually textbook neopositivism" (1986, 34). This turn to positivism "marks a definite break from his previously stated view," based on the notion of measurement disturbance (1986, 35; cf. Faye 1991, 186). According to Fine, "Bohr switched from [the] doctrine of actual physical disturbance to what one might call a doctrine of semantic disturbance" (1986, 35).

In their 1994 joint paper "Bohr's Response to EPR," Beller and Fine further elaborated this line of argument, not only tracing historical changes, but also analyzing what they see as an ensuing philosophical problem facing Bohr. According to the authors, Bohr originally disagreed with Heisenberg's operational approach to the meaning of concepts underlying his 1927 uncertainty paper. Specifically, he "objected [. . .] to Heisenberg's conflation of definition and observation," in favor of the complementary relation between the two. As a response to the EPR challenge, however, he came to adopt precisely the Heisenbergian operational approach, tracing the possibility of definition back to the possibility of measurement (Beller and Fine 1994, 18; cf. Beller 1999, 159, 184).[33] This raises, however, the following conceptual difficulty for Bohr. If, as he emphasizes, any observation involves an uncontrollable interaction between object and instrument, as exemplified by an uncontrollable exchange of momentum in the case of space coordination, this means that such an exchange "cannot be measured" and therefore "cannot be assigned a clear meaning" (1994, 20f.). In Beller and Fine's account, from the operational viewpoint, the idea of the uncontrollable exchange of momentum or energy, "which is supposed to ground [Bohr's] physical picture of quantum uncertainty," thus becomes "problematic" (1994, 22).[34] In this way,

> [. . .] [the challenge of] EPR drives the concept of a measurement disturbance, the central ingredient in Bohr's philosophy of complementarity, onto the horns of a dilemma (Beller and Fine 1994, 23; cf. 27f.).[35]

---

[33] In this connection, Beller and Fine point to logical positivists', specifically Philipp Frank's enthusiastic reception of this move of Bohr, which was followed by a positive philosophical exchange between the two (1994, 19f.).

[34] As Beller puts it elsewhere, "operationalism is especially unsuited – in fact it undermines – Bohr's philosophy of complementarity" (1999, 152).

[35] In her book *Quantum Dialogue*, Beller argues that not only do "unbridgeable gaps exist between Bohr's pre-1935 and post-1935 philosophies," but also that "conflicting opinions of realism and positivism" are often both present at the same time (1999, 172; cf. 275). Specifically, his response to EPR contains two conflicting voices, "one voice rooted in the past (before 1935), the other emerging into the future (after 1935)" (1999, 146). That is, "[t]he old voice, holding to the notion of physical disturbance, brings this notion to a dead end," while "[t]he new, emerging operational voice will culminate in unreserved verificationism and a future repudiation of the notion of disturbance" (1999, 150f.).

That is, Bohr is faced with the incompatibility between the operational approach, required to counter the EPR argument, and the notion of "uncontrollable disturbance" as "the physical basis of quantum uncertainty" (1994, 23). Unable to solve this dilemma, "Bohr moved from detailed considerations of measurement interactions and disturbances to broad discussions of the 'general epistemological lessons' of quantum theory." In this way, "[a]s a result of EPR, Bohr eventually turned from his original concept of disturbance, to make a final – and somewhat forced – landing in positivism" (1994, 29; cf. Beller 1999, 150).

It is here worth touching briefly on Carlsten Held's account of Bohr's philosophical turn, offered in his 1993 paper "The Meaning of Complementarity." Like Fine and Beller, Held also maintains that, as a result of his debate with EPR, Bohr shifted from a realist to a non-realist position (1994, 892f.; see PWNB, 4:7, 13). In contrast to the former two, however, Held conceives this philosophical transition not as a drift into a dilemma, but as an advance toward "logical consistency" (1994, 879). In his view, Bohr's complementarity as it was originally introduced was, as Folse argues, "fundamentally realist" (Held 1994, 891). Because of this realism, however, his thought was also "paradoxical" and "inconsistent" in character, specifically with regard to wave-particle duality (1994, 872).[36] For, under realist presuppositions, wave-particle duality, "if it is in fact a proper example of complementarity," shows that not only the experimental setups, but also the object's properties themselves are mutually exclusive (1994, 875). From the mid-1930s onward, however, Bohr strove to develop another notion of complementarity, which, breaking with the idea of wave-particle complementarity, "no longer refers to actual descriptions of atomic objects but to incompatible observables" (1994, 872, 880). According to Held, this mature conception of complementarity is philosophically "non-realist" and logically free of contradiction (1994, 892; cf. 879, 891).[37]

In the present section, closely linked with the previous one, I have surveyed several major studies on possible diachronic changes in Bohr's philosophical thought, showing how they are divided over whether and in what way such alterations occurred. There appears, however, to be at least one crucial notion common to most commentators: They presuppose that such alterations in Bohr's thought, if any, are to be described along the linear axis stretching between realism and anti-realism. Our review in the previous and the present sections taken together, we can indeed say that many prior studies of Bohr's complementarity confine themselves within this realism/anti-realism framework. While conflicting with each other, they share the tacit assumption that Bohr's philosophical orientation in a given period may be

[36] According to Held, although Bohr "alludes to dialectics" in his early argument for an inevitable "revision of our fundamental concepts" and even "logical principles," from the start he *strove* for logical consistency without immediately achieving it (1994, 879).

[37] Held also points out that, for the later Bohr, the logical consistency of complementarity is secured "by the mathematical consistency of the formalism of quantum mechanics" (PWNB, 3:25; Held 1994, 879). In his book *Die Bohr-Einstein-Debatte* (1999), Held re-situates his account of Bohr's complementarity within the broader context of the relation between the two physicists' approaches to quantum physics and philosophy.

identified with a specific position such as realism, anti-realism, or some intermediate or mixed standpoint. I wish to raise the following questions, however: Is it not rather the case that his idea of complementarity – at least in the 'early' and 'middle' periods – goes beyond the linear axis of realism/anti-realism? Does complementarity not pertain to the very *relation* between such conflicting philosophical positions, or their relation of conflict itself, instead of simply subscribing to one (or a mere mixture) of those positions? If this is the case, we must approach Bohr's complementarity quite differently than in the prior studies reviewed above. In particular, the issue of the historical development of Bohr's thought may also then require more complex considerations. In fact, I will seek to show in Chapter 4 that while Bohr's philosophical orientation did change significantly, this process cannot be seen as a transition along the axis of realism/anti-realism, but should be reinterpreted from a broader philosophical perspective yet to be developed. Before setting out such an analysis of my own, however, I wish to survey in the next section yet another thematic complex of prior interpretations: the relation of Bohr's thought to the history of modern philosophy.

3.4 Complementarity in the History of Modern Philosophy

Many prior studies of Bohr's thought have concerned themselves with the question of how to situate his idea of complementarity within the history of modern Western philosophy. Specifically, the philosophers Kant, Kierkegaard, James, and Høffding, among others, are considered by some commentators to have influenced Bohr or offered a conceptual background for his ideas (see Honner 1987, 73). According to Max Jammer, for example, the Danish philosopher Søren Kierkegaard's thought had an impact on the young Bohr, mediated by the latter's philosophical mentor Harald Høffding. Kierkegaard's "qualitative dialectic," and specifically his concept of the "leap," which were expounded favorably by Høffding, may have served as a source of Bohr's idea of quantum discontinuity (1989, 179f.; cf. Faye 1991, 37). Jammer also points out that "Bohr was strongly influenced by William James" as well, again "probably through Høffding" (1989, 182). As is suggested by his complementarity argument with regard to psychology, Bohr's complementarity seems to have been inspired by James's work, notably his account of the introspection of one's "stream of thought" in the *Principles of Psychology* (Jammer 1989, 184).[38] Further, according to some commentators, Høffding not only played an intermediary role, but his

[38] In his last interview, conducted by Kuhn and Petersen, Bohr testified that, in his youth, he had read with fascination James's work, specifically the chapter "The Stream of Thought" of the book just mentioned. *Archive for the History of Quantum Physics*, p. 6. With reference to this and other documentary sources, Klaus Michael Meyer-Abich analyzed the philosophical connections between Bohr's complementarity and James's psychological thought (1965, 133ff.). Further, drawing partly on Jammer's and Meyer-Abich's studies, Gerald Holton discusses extensively Bohr's debt to the thought of James, Kierkegaard, and Høffding. Gerald Holton, "Roots of Complementarity," in Holton (1973, 115–61). See also Folse (1985, 43–49) and Murdoch (1987, 225–29).

own philosophy left its imprint on Bohr's thought. Jan Faye, in particular, extensively studied the Bohr–Høffding relation, spelling out an essential influence of Høffding's views in epistemology, the philosophy of mind, and other related fields (1991, Part I, 1–109; cf. Kaiser 1992, 231ff.). It should be noted, however, that the claims of these studies on philosophers' direct or indirect *influences* on Bohr are not accepted universally, but contested or rejected by some on various grounds such as insufficient documentary evidence.[39] Although these conflicting views on the above philosophers' possible role are of historical and biographical interest, I will not enter further into them here.

On the other hand, Bohr's possible relation to Immanuel Kant's or Kantian philosophy has been addressed in a different manner. Largely admitting the lack of evidence for a direct influence, commentators rather focus on the question as to whether Bohr's complementarity bears conceptual resemblance to Kantian philosophy, and/or whether it historically emerged from an intellectual background formed by the latter. In his 1971 book *Die Einheit der Natur* [*The Unity of Nature*], von Weizsäcker pointed to philosophical affinities between Kant's and Bohr's thought, noting that "Niels Bohr is the only physicist in our time who – as far as I know, without having been influenced by Kant – proceeded from fundamental insights similar to Kant's" (1971, 424/342f.). These insights are characterized by the key Kantian term "transcendental" as against both "transcendent" and "empirical." That is, Bohr primarily concerns himself, not with "metaphysical hypotheses" nor with "particular experiences," but with "the preconditions of the possibility of experience." It is only from this point of view that one is "able to do justice to Bohr's doctrine of the indispensability of classical concepts" (1971, 426/345; cf. Honner 1987, 10). Specifically, "Bohr's dichotomy of space-time description and causality corresponds to Kant's dichotomy of the forms of intuition and the categories (and principles) of the understanding" (1971, 228/185, trans. mod.). Further, in von Weizsäcker's account, even though Bohr's position might seem to be akin to Machian positivism, he differs from Mach not least in maintaining that "'phenomena' are always 'phenomena involving things (*Phänomene an Dingen*),'" and that "the true role of things is that they are not 'behind' but 'in' the phenomena" (1971, 228/185; cf. Honner 1987, 15). This "comes very close to Kant's view that the concept of the object is a condition of the possibility of experience."

Generally speaking, however, commentators are divided over the possible philosophical links between Kant and Bohr. John Honner basically follows von Weizsäcker in regarding Bohr's complementarity argument as "transcendental" in the Kantian sense that it concerns "a necessary condition of the possibility of ordering experience" (Honner 1987, 105; cf. Beller 1999, 205).[40] Clifford A. Hooker also

---

[39] Against Jammer's and others' analyses, David Favrholdt seeks to refute the "myths" of the influence of philosophers (specifically Kierkegaard, James, and Høffding) on Bohr's thought (Favrholdt 1992, esp. 42ff.). Abraham Pais, the biographer of Bohr, is also negative about the role of philosophers in his "discovery of complementarity" (1991, 424).

[40] Honner uses the term "transcendental" to "signify a concern with the necessary conditions of the possibility of experiential knowledge, which includes unambiguous reports of that knowledge"

considers Bohr's conception of knowledge to be "a modification of Kant's in the light of contemporary experience" (1994, 184). Specifically, in Hooker's account, Bohr's thought represents a challenge to "Newtonian classical physical intelligibility" as defended by Einstein, on the basis of an alternative Kantian conception of intelligibility (1994, 182).[41] In contrast, Henry Folse rejects the Kant-Bohr connection in such regards. To be sure, admits Folse, "Bohr's restriction of the classical concepts to the description of phenomenal objects has a certain Kant-like appearance," in that "it issues in statements about the logical requirements for the proper use of concepts in describing nature." Yet, he contends, such an appearance is deceptive, because Bohr's claims "have nothing to do with *how* experienced phenomena arise in the subject's consciousness, as the Kantian statements do, but have to do only with communicating a *description* of a phenomenon as an objective datum already given in experience."[42] For this reason – and in accord with his realist reading of Bohr – Folse claims that "Bohr's talk about the use of concepts is tellingly non-Kantian" (1985, 219; cf. Honner 1987, 12). He gives yet another reason for dissociating Kant's and Bohr's thought: the latter's opposition to a priorism. That is, "Bohr based his whole argument for the correct use of classical concepts on a *physical* discovery, a purely *contingent* fact," while "a true Kantian approach would never argue [...] that such a physical discovery would demand a change" in the proper use of concepts (1985, 218; cf. Kaiser 1992, 223).[43]

In his 1992 paper "More Roots of Complementarity," David Kaiser presented an argument that included new or newly articulated points of analogy between Bohr's complementarity and Kant's philosophy.[44] According to Kaiser, while Folse rightly notes that Bohr rejects Kant's a priorism, this cannot be made a ground for altogether denying their philosophical affinity (1992, 222f.).[45] First – as von Weizsäcker and

(Honner 1987, 13). For a critique of his view of Bohr's complementarity as transcendental, see Plotnitsky (1994, 78f.).

[41] See Faye and Folse (1998, 15). Paul K. Feyerabend points to a parallel between Kant's and Bohr's views by saying that "[l]ike Kant before him, Bohr observes that our experimental statements are always formulated with the help of certain theoretical terms and that the elimination of these terms would lead [...] to complete chaos" (1981, 88). Murdoch also agrees, with some reservations, that there are some "Kantian elements in Bohr's philosophy" (1987, 229). In his article "Physical and Philosophical Issues in the Bohr-Einstein Debate," Abner Shimony also makes a brief "comparison of Bohr with Kant" (Feshbach et al. 1988, 285–303, on 299).

[42] Here I refrain from critically assessing Folse's understanding of Kant's thought as presupposed in his negative account of the Kant–Bohr relationship.

[43] As a support of his argument against Kant–Bohr connections, Folse also points to the fact that Bohr was influenced by the anti-Kantian philosopher James's approach to the description of experience (1985, 49; cf. Kaiser 1992, 223).

[44] Kaiser starts by noting that "Bohr describe[s] complementarity with language highly reminiscent of Kant's," exemplified by the use of expressions such as "forms of perception" and "modes of perception" (Kaiser 1992, 222). See PWNB, 1:1 and 5.

[45] It is noteworthy that the Neo-Kantian development of transcendental philosophy from the late nineteenth to the early twentieth century had the tendency to modify Kant's a priorism. Ernst Cassirer, in particular, sought to reconcile Kantian transcendentalism with the new conceptual situation of physical science by reinterpreting such particular forms of cognition as

others have already pointed out – "Bohr, like Kant, advocated a two-faculty epistemology," corresponding to the distinction between the forms of intuition and the understanding (1992, 222; see PWNB, 2:65). Second, in a draft of his Como lecture, he made "a distinction between 'closed systems of objects' and 'phenomena'" in parallel with Kant's noumena/phenomena distinction (Kaiser 1992, 227). The third point, particularly relevant to my later discussion, is that Bohr shared with Kant what Kaiser terms "conceptual containment" as a specific operation of transcendental thought (1992, 230). In the *Kritik der reinen Vernunft* [*Critique of Pure Reason*], Kant notes that "[i]f we *add* to the concept of the subject of a judgment *the limitation under which the judgment is made*, the judgment is then unconditionally valid."[46] Kaiser designates this kind of mechanism – that of "including within the concept itself the conditions and limitations under which the concept is made" – as "conceptual containment" (1992, 219f.). According to Kaiser, this mechanism, "together with the limitation of knowledge to objects of possible experience," generally "provide[s] Kant's solution to the problems of metaphysics."[47] He argues that the same mechanism of conceptual containment is at work in Bohr's complementarity as well, specifically in his response to EPR. Bohr's essential point of countering the EPR argument is that information regarding the "experimental arrangements and procedures," that is, "the conditions and limitations under which each individual judgment is made," must be included in the judgments in order for them to be valid (1992, 230). The same line of thought may be more manifestly seen in Bohr's later remarks, such as the one quoted in Section 2.4 of this study that "objective description can be achieved only by including in the account of the phenomena explicit reference to the experimental conditions" (PWNB, 3:12).[48] In this way, concludes Kaiser, the mechanism of conceptual containment, along with other factors, forms an essential link connecting Kant's transcendental philosophy and Bohr's complementarity.[49]

Euclidian space-time and deterministic causality not as historically immutable, but as subject to revision. This view, however, differs from Bohr's, according to which classical concepts remain indispensable in the sense that all experience is ultimately expressed in terms of them (see von Weizsäcker 1971, 424/343). See Ernst Cassirer, "Zur Einsteinschen Relativitätstheorie: Erkenntinistheoretische Betrachtungen," and "Determinismus und Indeterminismus in der modernen Physik: Historische und systematische Studien zum Kausalproblem," in Cassirer (1977, 1–125, 127–397).

[46] Kant (1781, A27/B43; trans., p. 72) (emphasis by Kaiser in 1992, 218f.). As is customary in citations from the *Kritik der reinen Vernunft*, I designate the first and the second editions of the book as A and B, respectively, giving the page numbers in the two editions with a slash in between.

[47] Conversely, in Kant's view, "uncontained concepts" give rise to illusion (Kaiser 1992, 220). For example, "if I ascribe redness to the rose in itself, or extension to all objects in themselves, without paying regard to the determinate relation of these objects to the subject, and without limiting my judgment to that relation, illusion then first arises" (Kant 1781, B70; trans., p. 89).

[48] Kaiser himself quotes a similar passage from an unpublished manuscript of Bohr (Kaiser 1992, 230).

[49] Kaiser suggests that the Kantian aspects of Bohr's complementarity resulted from Høffding's intermediary role, through which Bohr was exposed to Kant's philosophy (1992, 231ff.).

Catherine Chevalley also traces Bohr's complementarity back to the Kantian tradition, but in a manner more substantively engaged with the *history* of philosophy. In her 1994 article "Niels Bohr's Words and the Atlantis of Kantianism" and other related works, Chevalley focuses on Bohr's distinction between *Anschauung* (intuition) and symbol. In her account, Bohr holds that "the language of quantum formalism is a '*symbolic scheme*' which is in deep contrast to the 'intuitive description' of physical quantities provided by classical concepts" (1994, 34).[50] Noting that, in his terminology, the English word 'perception' as in the phrase 'forms of perception' corresponds to the German '*Anschauung*' (see 1994, 39),[51] she summarizes Bohr's view: "'concepts' or 'pictures' entail *Anschaulichkeit* and the objectivity of classical physics," while "'symbols' implies no *Anschaulichkeit* and a new kind of objectivity in [. . .] quantum physics" (1994, 36). Chevalley argues that this "contrast between '*Anschauung*' and '*Symbol*'" goes back to the Kantian philosophical tradition (1994, 48). In the *Kritik der Urteilskraft* [*Critique of Judgment*], Kant made a distinction between two kinds of 'hypotyposes' or presentations, "one 'schematic' and the other 'symbolic'" (Chevalley 1994, 43).[52] That is, "in Schematism we enjoy a direct presentation of the concept in intuition, while in Symbolism we deal with an indirect presentation" (1994, 44; cf. 1995, 24–26). This Kantian distinction was taken over and reconfigured by subsequent (mostly German) thinkers from Goethe through Helmholtz to Cassirer in such a way as to extend the notion of symbol to all kinds of experience (1994, 45–48; cf. 1995, 28f.). According to Chevalley, the Kantian tradition of the schema/symbol distinction with the post-Kantian progressive emphasis on the symbol constitutes the context within which Bohr's thought emerged.[53] In this way, he "construct[ed] a new kind of epistemology" beyond both "Kantian foundationalism and naïve empiricism" (1995, 32).[54]

So far in this section, we have seen various studies of Bohr's relation to the thought of *earlier* philosophers, especially of Kant.[55] The relevance of philosophers

[50] Chevalley traces the pre-1927 development of Bohr's thought from the introduction of "formal analogy" around 1920, through its development into the notion of "symbolic analogy" in 1925, until the emergence of complementarity (1994, 36f.; 1995, 15–17).

[51] As comparison between the English and the German versions of Bohr's papers indicates, both the phrases 'forms of perception' and "modes of perception" (PWNB, 1:90) correspond to "*Anschauungsformen.*" See, for example, the German version of his Como paper, "Das Quantenpostulat und die neuere Entwicklung der Atomistik," *Naturwissenschaften* 16 (1928): 245–57; reprinted in Bohr (1985, 156–83, on 183).

[52] See Kant (1790, 255) (in the original edition).

[53] In Chevalley's account, it is not that Bohr simply followed the tradition, but that he "criticize[d] from the inside the former philosophical language that he inherited, [. . .] ben[t] it, so to speak, against itself" (1995, 12). She designates this process as a "destructuration of the theoretical language" (1995, 14; cf. 24).

[54] For another approach to Bohr's thought with an emphasis on its Kantian, post- and neo-Kantian background, see Brock (2003), which contains favorable but partly critical comments on Chevalley's analysis (2003, 136–39).

[55] Some commentators have also pointed to affinities of Bohr's complementarity with Hegelian dialectics. Heisenberg associated "quantum-theoretic complementarity" with the "Hegelian triad" of thesis, antithesis, and synthesis (1969, 285). In Rosenfeld's account, "[s]ince his early youth, Bohr had been preoccupied by [the] problem of the ambiguity of language, and had with sure

contemporaneous with Bohr, however, has also at times been pointed out. As already mentioned (in footnotes in Sections 2.3, 3.2, and 3.3), some commentators have suggested a possible influence of logical positivism on Bohr, particularly around the crucial time of his debate with EPR.[56] In their 1998 joint "Introduction" to *Philosophical Writings of Niels Bohr*, Vol. 4, Faye and Folse indicate that, during the mid-1930s, Bohr corresponded with the leading logical positivist Otto Neurath, who perceived a kinship between logical positivism and complementarity and yet was critical of Bohr's 'metaphysical' mode of expression. According to Faye and Folse, this critique by Neurath may be seen as having "bor[n] fruit" in Bohr's reaction to EPR, in which "he gave up entirely on any attempt to use classical descriptive terminology to refer to a physical world in-itself" (1998, 8f.). During the same period, he also had close contact with Philipp Frank, "another leading member of the Vienna Circle," who "attributed to Bohr a positivistic view of physical reality whereas to Einstein a purely metaphysical view."[57] Through this communication with logical positivists, Bohr eventually shared with them "the view that physical reality could not be meaningfully referred to as something existing behind the phenomena" (PWNB, 4:9).[58]

In this final section of Chapter 3, we have reviewed a number of studies that seek to situate Bohr's complementarity in the history of modern philosophy and to associate his thought with a wide range of philosophical approaches.[59] While some of these interpretations are marked by their insightful suggestions, they generally remain to be examined with regard to their validity and scope of implications. Proper understanding of Bohr's relation to various philosophers' ideas, however,

---

intuition grasped its essentially dialectical character." Specifically, in reference to Møller's *The Adventures of a Danish Student*, which inspired Bohr a great deal, Rosenfeld characterizes "the quandaries of the licentiate lost in the maze of self-reflection" as a "lesson in hegelian logic" (1967, 121). On Rosenberg's Marxist interpretation of complementarity as a form of "dialectical materialism," see Jacobsen (2007, esp. 12f.). Bohr himself at times used the terms "dialectic" and "dialectics," though not technically and not specifically in the sense of Hegelian or Marxian dialectics (NBCW, 7:335; PWNB, 2:63; see Petersen 1963, 303). For an inquiry into the relation between quantum mechanics and Hegelian philosophy, see Pitt (1971).

[56] Jammer notes that logical positivism, specifically its doctrine of verificationism, "became influential among physicists in the twenties" (1989, 185).

[57] According to Faye and Folse, this series of intellectual interactions between Bohr and logical positivists led, in particular, to his participation in the Second International Congress for the Unity of Science held in Copenhagen in 1936, which was organized by Neurath and the Danish positivist Jørgen Jørgensen (Faye and Folse 1998, 7). At this conference, Bohr delivered an address entitled "Causality and Complementarity" (included in PWNB, 4:83–91). Röseberg, another commentator on Bohr and logical positivists, stresses, however, that, despite such interactions, the two sides never converged in their philosophical views, and that there remained, especially on the part of the philosophers, "misunderstandings and misinterpretations of Bohr's ideas" (1995, 117).

[58] It is worth noting that this account by Faye and Folse seems to deviate significantly from the latter's earlier realist reading of Bohr as surveyed in Section 3.2. Here I do not, however, enter into the meta-interpretive question of possible historical changes in Folse's (or any other's) interpretation.

[59] I have not thematized the influence of Bohr's complementarity *on* other scientists or thinkers. For an account of the Japanese physicist Yoshio Nishina's wartime political use of the concept of complementarity, see Ito (2002).

will naturally rest on a substantive analysis of his idea of complementarity itself. In order to assess his possible connections with Kantian philosophy, in particular, we must elaborate the epistemological dimension of his complementarity argument. In the next chapter, I will accordingly set out my own interpretive approach to Bohr's complementarity, with a focus on both its structural complexity and its historical path of development.

Chapter 4
A Philosophical-Historical Analysis of Complementarity

In the previous chapter, we have seen a number of prior interpretations of Bohr's complementarity which are widely different from – and some of them diametrically opposed to – each other. To be sure, it is quite common that commentators are divided on any given philosopher's or scientist's ideas. Yet the degree of divergence in the reading of Bohr appears to exceed those in the case of most philosophers as well as many other modern physicists, not least Einstein or Heisenberg. In my view, this lack of consensus largely derives from the nature of the subject matter, Bohr's philosophical thought itself. This is not to say, however – and it would be too facile to conclude – that his thought is simply confused or incoherent. Rather, the following set of questions may be posed: Did Bohr's idea of complementarity not undergo subtle yet significant changes over time, changes that he himself never explicitly acknowledged? Even within one and the same period, is it not structurally too complex to be reduced to a single conceptual framework? Does this complexity in the structural and historical dimensions of his thought not go beyond the conventional interpretive schemes, specifically the opposition of realism and anti-realism?

It is from these points of view that, in the present chapter, I will offer a detailed analysis of Bohr's complementarity. Taking seriously his metaphor of 'spectators' and 'actors,' in particular, I wish to spell out both the structural complexity and the historical development of his thought in terms of the 'spectator–actor' relationship. In Section 4.1, examining Bohr's 'early' complementarity argument, I introduce the distinction between two – static (or more specifically, static-contrastive) and dynamic – conceptions of complementarity, whose philosophical implications are further explored in the second section. Section 4.3 focuses on his reorganization of complementarity in the 'middle' period, which is characterized by an extension of the dynamic conception from non-physical fields to quantum theory itself and yet, paradoxically, also by the formation of another static idea, the static-symmetrical conception. Finally, in Section 4.4, I address the way in which this static-symmetrical conception led to the development of his objectivist philosophy in the 'late' period.

M. Katsumori, *Niels Bohr's Complementarity*, Boston Studies in the Philosophy of Science 286, DOI 10.1007/978-94-007-1748-0_4,

4.1 The 'Early' Period: Two Conceptions of Complementarity

In this section, I explore Bohr's 'early,' pre-EPR formulation of complementarity as it was outlined in the first two sections of Chapter 2. In so doing, I wish to pay special attention to the relation between complementarity in quantum theory and that in other fields of knowledge and experience. Contrary to the view held by some commentators that Bohr's complementarity argument in non-physical fields is nothing more than "incomplete" and "vague" analogies (Beller 1999, 264; cf. 270), I consider that his argument regarding those areas offers essential clues for understanding the overall conceptual structure of complementarity.[1]

In Chapter 2, we saw that Bohr, at least in the 'early' period, conceives complementarity as a relation of contrast between the roles of 'actors' and 'spectators,' citing the dictum that 'we are both spectators (or onlookers) and actors in the great drama of existence.' Before proceeding to complementarity in fields outside physics, let us first look back on complementarity in quantum theory to analyze it in terms of the 'spectator–actor' relationship. The crux of Bohr's 'early' complementarity argument with regard to this field may be most briefly rendered as follows: According to quantum theory, any observation of atomic phenomena involves an unavoidable and uncontrollable "interaction with the agency of observation." Therefore, when observation is carried out, "an unambiguous definition of the state of the system is naturally no longer possible, and there can be no question of causality." On the other hand, the definition of the state of the system "claims the elimination of all external disturbances," in which case "any observation will be impossible, and [...] the concepts of space and time lose their immediate sense." Faced with this apparent dilemma, one must regard the space-time coordination and the claim of causality as "complementary" features of the description (PWNB, 1:54f.).

We can readily see how this complementarity idea in quantum theory may be characterized in terms of the roles of 'actors' and 'spectators.' Space-time coordination by means of observation, which involves an object–agency interaction, has the character of being an 'actor.' On the other hand, the claim of causality, associated with the definition of the state, which excludes such an interaction, has the mode of being a 'spectator.' In the former case, the subject behaves as an 'actor' in its relation to the object, while in the latter it behaves as a 'spectator' and thus leaves the object intact, maintaining the separation of the two sides.[2] Bohr conceives these two cases, corresponding to the roles of 'actor' and 'spectator,' as complementary in the double

[1] Bohr himself often claimed that he was "not dealing with more or less vague analogies, but with an investigation of the conditions for the proper use of our conceptual means of expression" (PWNB, 2:2). Unlike Bohr himself, however, I will attend not only to the commonalities, but also to the differences and tensions between his notions of complementarity in different fields.

[2] Bohr obviously holds that the agency of observation or measurement belongs to the side of the observing *subject*. Otherwise his entire attempt to connect complementarity in quantum theory with the epistemological problem of the subject–object relation would be incomprehensible.

sense of 'mutually exclusive' and 'jointly completing.' This 'actor–spectator' complementarity can be characterized in less figurative terms as the relation between *involvement* in the phenomenon and *detachment* from it. The two complementary relata – the 'actor's' involvement and the 'spectator's' detachment – may also be expressed symbolically as S∼O and S/O, respectively, in terms of the interaction (∼) and the separation (/) of the subject S and the object O. It is worth stressing that the expression S∼O does not imply, as it might seem, that S and O exist independently in the first place and subsequently come into interaction, but rather designates the situation in which the two sides are interdependent from the outset. In other words, while the 'spectator's' detachment conforms to the dichotomy of subject and object, the 'actor's' involvement tends to dissolve this dichotomous framework. It should also be noted that what Bohr means – and here I also mean – by the 'actor' is, notwithstanding its designation, not simply a free intentional subject of action, but rather a subject that finds itself in *interaction* with the object, where the alternative between activity and passivity no longer holds.

This first step of my analysis of Bohr's complementarity may be compared with, and has in fact been partly inspired by, Klaus Meyer-Abich's somewhat similar approach to the same subject matter. In Meyer-Abich's account, "[t]he space-time description of atomic systems, since it is possible only through observation, takes place [...] in the form of relational statements S′//S/O," while the claim of causality is "expressed in the form S′/O," where S denotes the observing subject and S′ the physicist as a theorizing subject (1965, 153). Here – and further with regard to fields outside physics as well – he focuses on different (single and double) structures of subject–object relations that are regarded as complementary to each other. As we can see, however, Meyer-Abich makes no basic distinction between the ways in which subject and object are related in each case, but rather generally presupposes the dichotomy of subject and object. This is illustrated by his use of one and the same symbol (/) to denote all subject–object relations. In other words, he pays little attention to *different modes* of the subject–object relation – such as involvement and detachment – which I consider to be crucial to Bohr's thought, and reduces these different modes to different combinations of dichotomies.

Let us now start to extend our scope to fields other than quantum theory, but for the moment add only one such field: biology. To this extent, the analogy between the two areas seems to be quite plain and straightforward. As we saw in Section 2.2, Bohr maintains that the physical analysis of life involves an unavoidable interaction with the organism, and, more specifically, that if we tried to carry such an analysis so far that "we could tell the part played by the single atoms in vital functions," we would doubtless kill the organism (PWNB, 2:9). This indicates that the physical analysis of the phenomena of life is analogous to the space-time observation of quantum phenomena, both involving an unavoidable interaction between the agency of observation and what is observed. The typical aspects of life, on the other hand, insofar as it is treated as *object* of investigation, correspond to "the applicability of the concept of atomic states" (PWNB, 1:23) associated with the claim of causality, since both require avoiding the disturbance of the phenomena. The analogy between the two fields may thus be expressed by Table 4.1.

Table 4.1 Complementarity in quantum theory and biology

	'Actor'	'Spectator'
Quantum theory	Space-time coordination (observation)	Claim of causality (state definition)
Biology	Physical analysis	Access to the typical aspects of life

The situation becomes more complex, however, once we take into consideration other fields of experience such as psychology and epistemology. As we have seen earlier, Bohr points out that an attempt to observe "the accompanying physiological processes" brings about "an essential alteration in the awareness of volition" (PWNB, 1:100f.). It might first appear sufficient to note that this physiological observation corresponds to the space-time observation of quantum phenomena and to the physical analysis of the phenomena of life. Yet its complement, "the awareness of volition" or "the feeling of free will" (PWNB, 1:100), can hardly be seen as an analogue of the definition of the state, because it does not rest on the separation of subject and object, whereas the definition of atomic states does. In other words, while it is as a 'spectator' that one defines atomic states, it is as an 'actor' that one feels one's free will. It is thus the feeling of free will or, more generally, the experience of psychical life that should be characterized as an 'actor's' involvement in the phenomenon. This being the case, what is complementary to psychical experience, namely, its physiological as well as introspective analysis, may be characterized as a 'spectator's' detached reflection on that experience. Similarly, as for complementarity in epistemology, one serves as an 'actor' when one immediately applies a concept, while one serves as a 'spectator' when consciously analyzing it. Further, looking back on biological problems, we may realize that there is a way of viewing complementarity in this field, too, as analogous to that in psychology and epistemology. That is, if the typical aspects of life are not taken as an object of knowledge, but as something immediately experienced, this experience – just as the experience of psychical life – constitutes a mode of involvement in the phenomenon. To sum up, the analogy between the fields of biology, psychology, and epistemology can for the moment be expressed by Table 4.2.

Comparison of Tables 4.1 and 4.2 indicates, however, an apparently odd state of affairs. As regards the field of biology, the physical analysis of life is characterized in Table 4.1 as an 'actor's' involvement and thus put into the left column, while in Table 4.2 it is classified in the right column, the column for the 'spectator's'

Table 4.2 Complementarity in non-physical fields (provisional)

	'Actor'	'Spectator'
Biology	Access to the typical aspects of life	Physical analysis
Psychology	Psychical experience	Reflective analysis
Epistemology	Application of concepts	Conscious analysis

Table 4.3 Complementarity in quantum theory and non-physical fields?

	'Actor'	'Spectator'
Quantum theory	Space-time coordination (observation)	Claim of causality (state definition)
Biology	Physical analysis	Access to the typical aspects of life
Psychology	Reflective analysis	Psychical experience
Epistemology	Conscious analysis	Application of concepts
	'Spectator'	'Actor'

detached reflection. Correspondingly, "access to the typical aspects of life" appears on the right side in Table 4.1 and on the left in Table 4.2. That is, the two tables do not seem to be consistent with each other, but to be separated by a kind of inversion of complementary relata. If we nevertheless attempted to combine the two tables directly, we would obtain a 'twisted' arrangement shown by Table 4.3. This situation of inversion not only remains latent in Bohr's texts, but perhaps was not even noticed as such by the author himself or by many commentators. In my view, however, awareness of this difficulty in connecting complementary relations in different fields – or of the impossibility of their straightforward connection – is indispensable for proceeding further to the essential points. In what follows, I will accordingly go on to examine the reason why the above structural peculiarity has arisen.

As we have seen so far, both Tables 4.1 and 4.2 concern complementary relations between involvement and detachment, between the roles of 'spectators' and 'actors.' On closer inspection, however, the right column of Table 4.2 proves not simply detachment, but detached *reflection*, and this suggests a structural difference between the states of affairs expressed by the two tables. As is exemplified by the case of introspection, where one reflects on one's own psychical experience, reflection does not stand side by side with involvement, but takes place as reflection *upon* involvement. We may therefore express the complementary relata in Table 4.2, the 'actor's' involvement and the 'spectator's' reflection, as S~O and $S'/(S{\sim}O)'$, respectively, where S' denotes the reflecting subject as distinct from the subject involved in the phenomenon, S. The distinction between S and S' is not substantial, however, but only functional, so that they can both refer to the 'same' person, for example. The expression $(S{\sim}O)'$ designates the involvement as viewed by the reflecting subject, and the prime in the expression signifies that S~O may be altered by the very act of reflection. This influence or disturbance by reflection will turn out to be of crucial significance to the structure of complementarity.

In Bohr's view, the 'spectator's' detached reflection, while contrasting and conflicting with involvement, proves not to leave the phenomenon purely intact, but to alter it, that is, to have itself an aspect of involvement. As we have seen earlier, he stresses that the physical analysis of a living organism entails an interference with it, that the introspection of one's psychical experience unavoidably influences it, and that the conscious analysis of a concept interacts with its very meaning – a set of situations analogous to "the conditions responsible for the failure of causality in the analysis of atomic phenomena" (PWNB, 1:23f.). This is the reason why, in the field of biology, the physical analysis of life, which is characterized as a 'spectator's' detached reflection in Table 4.2, is classified as an 'actor's' involvement in Table 4.1. This inversion by no means constitutes a simple contradiction or confusion, but derives precisely from the above paradoxical feature of reflection – detached reflection which, in being itself, turns out to have the character of its opposite, involvement. Since this is also the case with the reflective analysis of psychical experience and the analysis of concepts in the epistemological dimension, we could reasonably add these items – in parallel with the physical analysis of life – to the left column of Table 4.1, namely, the column for the 'actor.'

Our consideration so far enables us to see that Bohr's 'early' complementarity argument does not form a single conceptual framework, but consists of at least two distinct conceptions, which correspond to two possible readings of the dictum that 'we are both spectators and actors.' The first conception of complementarity, expressed by Table 4.1 alone, concerns the circumstance that one behaves as an 'actor' and as a 'spectator,' not at the same time or in the same situation, but only in separate and incompatible situations. That is, in this conception, the two complementary relata, exemplified by space-time coordination and the claim of causality, are mutually exclusive in the sense that only one or the other can be realized in a given situation. To be sure, in quantitative terms, both can be partially realized in inverse proportion to each other in accordance with Heisenberg's uncertainty relations. The point is, however, that the more one is involved in the phenomenon, the less one is detached from it – which might sound trivially true, but will prove not necessarily so. This conceptual scheme of complementarity, which is predominant in Bohr's 'early' account of quantum theory, may be characterized as *static and onefold*, that is, closed on itself as a two-term relationship. In other words, complementarity in this sense takes the form of a static *juxtaposition* of the 'actor's' involvement in the phenomenon, S∼O, and the 'spectator's' detachment from it, S/O.

In contrast, the second conception of complementarity – in the 'early' period largely limited to fields outside quantum theory – is expressed by Table 4.2 in conjunction with the left column of Table 4.1. This conception refers to the circumstance that when involvement in the phenomenon S∼O is reflected upon, this reflection $S'/(S\sim O)'$ proves to have itself a character of involvement, or, in other words, that being a 'spectator' itself turns out to contain the moment of being an 'actor.' The same above dictum – that 'we are both spectators and actors' – is interpreted differently here: We are 'spectators' and 'actors,' not separately, but in the same situation, or, to put it more aptly, *we are 'actors' precisely in being 'spectators.'* Specifically, as regards biology, physical analysis as a reflection on the phenomena of life turns out itself to be another involvement in those phenomena. As for psychology, the reflective analysis of psychical experience proves to be another instance of psychical experience. This state of affairs may be succinctly expressed by the following new, single table, Table 4.4 (instead of Table 4.3), which, apart from the field of quantum theory, represents an appropriate combination of the first two tables. Here, as can be seen, I have rewritten the phrase "access to the typical

Table 4.4 The dynamic conception of complementarity in non-physical fields

	'Actor'	'Spectator' / 'Actor'	
Biology	Involvement in life phenomena	Physical analysis / Involvement in life phenomena	...
Psychology	Psychical experience	Reflective analysis / Psychical experience	...
Epistemology	Application of concepts	Conscious analysis / Application of concepts	...

aspects of life" in the previous tables as "involvement in life phenomena," which seems to be a more rigorous analogue of the corresponding items in the other fields.

Further, in this conception of complementarity, the fact that reflection has itself a character of involvement becomes explicit, not for the subject engaged in that reflection, S′, but for a new subject reflecting on that reflection, say S″, where the distinction between S′ and S″ is, again, only functional. In other words, the fact that being a 'spectator' contains the moment of being an 'actor' is recognized only by a new 'spectator,' who reflects on what the first 'spectator' does. This second 'spectator' will in turn, however, on still further reflection, prove to have the character of being a new 'actor.' For this reason, one can never reach a final, unambiguous grasp of what is reflected upon, but rather undergoes a series of reflections that is in principle indefinitely extensible. This series of reflections, with alternating moments of detachment and involvement, can be schematically expressed as:

$$\mathrm{S} \sim \mathrm{O} \rightarrow \mathrm{S}'/(\mathrm{S} \sim \mathrm{O})' \rightarrow \mathrm{S}''/(\mathrm{S}' \sim (\mathrm{S} \sim \mathrm{O})')'' \rightarrow \ldots \text{[3]} \qquad (4.1)$$

This series of reflections may be illustrated by the passage from Møller's novel cited earlier, which describes how "I [...] divide myself into an infinite retrogressive sequence of 'I's'."[4] We can see that it is impossible to totalize the series, because

[3] This part of my analysis again partly draws on, but substantially recasts, Meyer-Abich's account of Bohr's complementarity with regard to epistemology. Meyer-Abich supposes the circumstance that "someone (S) knows something (O)," and then that this knowledge is described by another subject S′, and further that this description is understood by yet another subject S″. In his account, this process may be symbolically expressed as: S/O → S′//S/O → S″///S′//S/O → .... Here, again, as illustrated by his use of the same sign (/) for all subject–object relations, Meyer-Abich ignores the difference between involvement and detachment as distinct modes of the subject–object relation (1965, 173). Though less manifest, the same seems to apply to Aage Petersen's account of Bohr's complementarity (see 1963, 302f.).

[4] The remark by Møller's licentiate, quoted earlier in Section 2.2, that "the spectator anew becomes an actor" appears to be highly relevant to my discussion here.

any attempt to have in view the whole series will amount to the addition of yet another term to the series. It is also important to note that the influence exerted by reflection at each step cannot itself be unambiguously determined, that is, cannot be 'controlled.' For the determination of the interaction requires in its turn yet another reflection, which will introduce a new interaction with what is reflected upon, so that the phenomenon is uncontrollably displaced in meaning. This is presumably Bohr's reasoning when he emphasizes the inevitably ambiguous and "floating" character of our language. Bohr's second conception of complementarity, as analyzed so far, may thus be characterized as *dynamic and multifold displacement*. This dynamic conception of complementarity, while provisionally starting from the conventional subject–object scheme, nevertheless displaces and 'deconstructs' the latter.

Let us briefly consider the question as to whether the notion of 'mutual exclusion' applies to this second notion of complementarity as well. To be sure, since reflection presupposes what is reflected upon, the two complementary relata here, involvement and detached reflection, do not occur in separate situations, and are thus not mutually exclusive in the same sense as in the static structure considered earlier. Nevertheless, the concept of mutual exclusion will prove to be no less relevant to the dynamic conception. First, as we have seen, reflection carries with it an unavoidable interaction with involvement, and this implies that involvement, once reflected upon, no longer remains purely itself, but has become 'involvement as viewed by the reflecting subject.' In other words, involvement can never been grasped *as such*, so that 'involvement as such,' if it existed, would be incompatible with reflection on it. Second, we have also seen that detached reflection, once it is itself reflected upon, turns into its opposite, involvement, from the new reflective point of view. What is crucial here is that any given subject–object relation *as viewed from a given standpoint* is either involvement or detached reflection, and cannot be both simultaneously. For these reasons, as illustrated by Bohr's remark cited earlier on the conscious analysis of a concept, the 'spectator's' reflection "stands in a relation of exclusion" to the 'actor's' involvement (PWNB, 1:96). We can thus see that the notion of mutual exclusion is constitutive also of the dynamic conception of complementarity, but in a different way than in the case of the static one.

My analysis so far has shown that Bohr's 'early' idea of complementarity consists of two different conceptions – static-onefold and dynamic-multifold. These two conceptions of complementarity do not simply coexist in Bohr's thought, but stand in latent tension with each other. While the first conception rests on a contrastive juxtaposition of two items, the second tends to undermine this binary opposition, not by neutralizing, but by continually displacing the opposing items. These two heterogeneous, at least potentially conflicting modes of thought form what may be termed the double structure of Bohr's 'early' complementarity. In the next section, before proceeding to his thought in the 'middle' and 'late' periods, I will discuss – with reference to some of the prior studies reviewed in the previous chapter – further philosophical implications of his 'early' idea of complementarity as analyzed so far.

4.2 The 'Early' Period: Further Philosophical Implications

In Chapter 3, we reviewed various prior interpretations of Bohr's complementarity by physicists as well as historians and philosophers of science. In this section, I begin by taking up again some of the interpretive approaches and categories discussed there: (1) von Weizsäcker's account of complementarity, (2) interpretations along the axis of realism/anti-realism, and (3) the possible relevance of Kantian philosophy. In reference to these issues, I wish to elaborate further my interpretation of Bohr's 'early' idea of complementarity basically set forth in the foregoing section.

(1) My distinction introduced in the previous section between the two conceptions of complementarity may be reminiscent of Carl Friedrich von Weizsäcker's distinction between parallel and circular complementarity. As we saw in Section 3.1, von Weizsäcker's term "parallel complementarity" refers to a relation between "concepts of the same level" such as position and momentum, whereas "circular complementarity" involves a "circle of knowledge" such that "our knowledge has a priori presuppositions which nevertheless become corrected and reinterpreted retroactively by substantive findings obtained with their own help" (1976, 284, 294). This circular complementarity is exemplified by the relation "between space-time description and the claim of causality" and, outside physics, by the relation "between the definition (or analysis) of a concept and its immediate use" (1976, 293f.). It may now readily appear that what I have termed the static-onefold conception of complementarity is similar to von Weizsäcker's parallel complementarity, while the dynamic-multifold conception in my account resembles his notion of circular complementarity. This analogy indeed holds to a certain extent: The static conception of complementarity is characterized by a simple – and in a sense 'parallel' – juxtaposition of two items, while the dynamic conception, just as circular complementarity, goes beyond this juxtaposition and enacts a series of alternations between the complementary relata. Thus, from my interpretive point of view – despite the later Bohr's disagreement – von Weizsäcker's insight into a double structure of complementarity appears to be basically to the point.[5]

Nevertheless, my interpretation of Bohr's complementarity differs from von Weizsäcker's in the following respects. Since, in Section 3.1, I already pointed to his misunderstanding (admitted by himself) of the claim of causality as a description in terms of the Schrödinger wave function, here I wish to restrict myself to other, more properly *philosophical* differences between his and my readings. First, von Weizsäcker's interpretation is not concerned with the contrast between the roles of 'actors' and 'spectators,' or between involvement and detachment. Neither parallel nor circular complementarity in his account revolves around this conceptual pair, which is pivotal for my interpretation. Second, in von Weizsäcker's account

[5] Unlike many other commentators, Meyer-Abich is also favorable – even more favorable than I am – to von Weizsäcker's concept of circular complementarity, noting that it "applies [...] to Bohr's initial use of the concept of complementarity" in quantum theory as well as in psychology and epistemology (1965, 157).

of circular complementarity, although one of the complementary relata (the analysis of a concept, for example) modifies the other (the immediate understanding thereof), the character of each relatum is unambiguously determined and remains fixed. In contrast, my point is that the role of the 'spectator' turns out to have the character of an 'actor,' as is illustrated by the example that the analysis of a concept – or, more precisely, the analysis of the use thereof – proves to involve yet another use of the concept and/or a use of other concepts. This leads to a radical undecidability between the roles of 'spectators' and 'actors,' which is absent in von Weizsäcker's reading. In this way, my partition of Bohr's complementarity into static and dynamic conceptions is in a way analogous to, and yet irreducibly diverges from, von Weizsäcker's dual – parallel/circular – notion of complementarity.

(2) As we saw in Section 3.2, Bohr's philosophical position has been construed in widely different ways by different commentators: as subjectivist anti-realism by Popper, as realism by Folse, as "objective anti-realism" by Faye, and so forth. As I have noted, despite the sharp conflict with each other, many such interpretive accounts share the tacit premise that Bohr's philosophical orientation can be identified with a specific position along the axis of realism/anti-realism. My analysis in the previous section suggests, however, that, at least in the 'early' period, Bohr subscribes to neither realism nor anti-realism, nor any intermediate or simply mixed standpoint. Rather, he suspends the validity of both realism and anti-realism, and *situates the very relation between them within the framework of complementarity*. As indicated earlier, the detached 'spectator' posits an object as independent reality, while the involved 'actor' conceives the object as appearing only in its relation to him/herself. In the dynamic conception, in particular, the 'actor's' involvement in the phenomenon is realistically objectified by the reflecting 'spectator,' but this reflection subsequently proves to have itself a character of involvement, where the realist notion of an independent object breaks down. Since such an alternation of standpoints may proceed indefinitely, neither the realist nor the anti-realist position can fully and finally prevail over the other. This point, namely Bohr's double break with realism and anti-realism, which leads to the radical indeterminacy between them, appears to be missed by many commentators in their attempts to identify his philosophical position.

To be sure, there are a few exceptions among the commentators in this regard. Max Jammer, in particular, briefly notes that "Bohr avoided committing himself either to idealism or to realism," and that "for Bohr the very issue between realism and positivism (or between realism and idealism) was a matter subject to complementarity" (1974, 207).[6] This may indeed be regarded as similar to my above point on Bohr's attitude toward the relation between realism and anti-realism. Insofar, however, as Jammer here – like many others – has anti-realism represented by idealism or positivism, his view does not accord with mine. What I have rather

[6] As quoted in Section 3.2 (note 23), Murdoch makes a somewhat similar remark: The realist and the instrumentalist components of Bohr's philosophical thought are "so to speak, complementary sides of the phenomenon that is Bohr" (1987, 222). He does not, however, appear to use the term 'complementary' in any strict sense here.

provisionally called 'anti-realism' above does not imply the subjective construction of the world or the reduction of all beings to subjective ideas, whether empirical or a priori. Rather it primarily signifies the undoing of the dichotomy of subject and object, where the notion of the subject as an independent agency is no less undermined than that of the independent object. From this point of view, Bohr's complementarity may perhaps be better characterized – without using the term 'anti-realism' – as an undecidable suspension and alternation between the realist subject/object dichotomy and the 'non-realist' disruption thereof.

(3) In Section 3.4, we reviewed prior studies on the relation between Bohr's complementarity and the thought of various philosophers, particularly Kant's or Kantian philosophy. On the one hand, it is not in itself incorrect to characterize, as von Weizsäcker and Honner do, Bohr's philosophical approach as 'transcendental' in the broad sense of concern with the necessary conditions of the possibility of experience. Further, his contrast between space-time and causality, basic to the concept of complementarity in quantum theory, may also be likened to Kant's distinction between the forms of intuition and the categories. On the other hand, our consideration so far, specifically in the above subsection (2), indicates that Bohr's thought does not accord with transcendental *idealism* insofar as the latter fundamentally prioritizes the subjective principles of knowledge over that which is objectively known. In Kant's philosophy, the object of knowledge as a phenomenon obeys the a priori principles of knowledge which reside in the faculties of the knowing subject. By contrast, Bohr's notion of the 'actor's' involvement in the phenomenon, in particular, implies that the subject and the object are dependent on each other, and that the principles of knowledge reside, as it were, *between* the two sides.

As is well known, in the *Kritik der reinen Vernunft*, Kant characterizes his philosophical position in terms of the conceptual pair of "transcendental idealism" and "empirical realism." According to Kant, the "transcendental idealist," who regards all phenomena as "representations" and not as "things in themselves," may at the same time be an "empirical realist." Unlike the "transcendental realist," the empirical realist ascribes a reality to matter or "admit[s] the existence of matter without going outside his mere self-consciousness" (1781, A369f.; trans. pp. 345f.). In this way, despite their apparent opposition, transcendental idealism and empirical realism are fully compatible with each other, constituting two different aspects of one and the same philosophical position which is Kant's own. Here it is worth comparing this Kantian view with the above implication of Bohr's complementarity for the relation between realism and anti- or non-realism. The two thinkers' ideas here are indeed similar in that they both seek to combine apparently antithetical philosophical positions, a certain kind of realism and its obverse. Unlike Kant's approach, however, Bohr's complementarity treats the relation between realism and anti-realism – or rather the realist subject/object dichotomy and the non-realist undoing thereof – as not merely seemingly, but *actually conflictual*. As we have seen, in his static conception of complementarity, the two modes of the subject–object relation – dichotomy and interdependence – are mutually exclusive in the sense that they appear only in separate and incompatible situations. They are mutually exclusive in his dynamic conception as well, though in a different way: The subject/object dichotomy in detached reflection turns out itself to be a mode of its opposite, the interdependence of the two sides, for a new reflecting subject. We can thus see that neither the static nor the dynamic conception of complementarity implies a notion exactly equivalent to the Kantian relation between transcendental idealism and empirical realism.

In Section 3.4, I addressed, among other interpretations, Kaiser's account of the analogy between Bohr's complementarity and Kantian philosophy in terms of what he calls "conceptual containment." Kaiser maintains that, in Bohr's as well as Kant's thought, a concept or judgment is rendered valid by the mechanism of conceptual containment – that of including within the concept or judgment itself the conditions and limitations under which the concept or judgment is made – and that this mechanism serves as an essential link between the two thinkers' philosophical approaches. I basically agree with, and attach importance to, this interpretive account by Kaiser. Further, it appears to me that while, as Kaiser points out, Bohr's idea of conceptual containment is noticeable especially in his post-EPR work, the same idea has from the outset some relevance to his complementarity argument. In the understanding of this relevance, however, my reading of Bohr, and, by extension, my view of the Bohr–Kant relation, diverge from Kaiser's.

In my view, an aspect of the idea of conceptual containment is involved specifically in what I have termed the dynamic conception of complementarity, and this aspect is not a result of containment, but an ongoing process or movement thereof. As can be seen from my earlier discussion, in the dynamic conception, the 'spectator's' detached reflection on the 'actor's' involvement turns out not to be unconditionally valid, but to be bound by certain conditions and limitations that render the reflection itself another instance of involvement. When, for example, one analyzes one's own application of a concept, this analysis proves to be made only with the aid of another concept or set of concepts, and this condition – the one under which the analysis is carried out – renders the analysis another instance of application. This kind of recomprehension of reflection as another involvement from a new reflective viewpoint can be repeated indefinitely, so that the process of conceptual containment is never accomplished once and for all. In other words, Bohr's dynamic conception of complementarity implies that concepts and words, while subject to a movement of containment, *can never be fully contained* and instead every time exceed their temporary containment. This is the reason why, in Bohr's account, we can never attain purely unambiguous meanings of words in our use of language. In this way, while Kaiser's above point aptly focuses on an important aspect of complementarity, what is crucial to Bohr's thought, notably his dynamic conception of complementarity, lies in the fact that it implies the impossibility of full conceptual containment and thus a continual alternation between the moment of containment and the moment of surpassing thereof. Insofar as Kant's philosophy rests on the notion of full conceptual containment to ensure the possibility of universal and necessary knowledge, here we can see a key difference between the two thinkers' philosophical orientations.

So far in the present section, I have pursued philosophical implications of Bohr's 'early' complementarity with critical reference to several prior interpretations reviewed earlier. The remainder of this section will be devoted to further analysis of his thought with regard specifically to his dynamic conception of complementarity. In the previous section, we saw how, during the 'early' period, Bohr's complementarity as dynamic displacement operates in fields such as biology, psychology, and epistemology. Here there may arise the question: Is this dynamic conception not relevant to quantum theory, the field of Bohr's utmost preoccupation? As I will discuss in the next section, it is not until the mid-1930s that Bohr would carry over his dynamic conception in a fully significant manner into the field of quantum theory. In a certain sense, however, the dynamic conception is already from the outset relevant to all fields of knowledge and experience, including quantum physics. As suggested in Section 2.2, the relation between complementarity in quantum theory and that in epistemology is not only analogical, but also internal, for the simple reason that quantum theory is one of the "domains of knowledge" in which concepts and words are used and defined. When, for example, Bohr *speaks* of "space-time coordination and the claim of causality" as "complementary [...] features of the description" (PWNB, 1:54f.), we can analyze, as Meyer-Abich attempts to do, how this statement on complementarity is itself subject to complementarity in the epistemological dimension (see Meyer-Abich 1965, 158). That is, Bohr's statement may be regarded as reflection on both space-time coordination (S∼O) and the claim of causality (S/O), thus assuming the form S′/((S∼O) vs. (S/O))′.[7] Further reflection of this kind will, as we have seen earlier with regard to the dynamic conception in general, lead to an indefinitely extensible series in which the meanings of the concepts used and defined are each time uncontrollably displaced. This dynamic situation with regard to quantum-theoretical knowledge or discourse, though not elaborated by Bohr himself, appears to form a background of his view that one can never attain fully unambiguous meanings of words even in natural science, hardly any more than in literature or religion. This internal epistemological relevance of the dynamic conception to quantum theory as well as to any other field seems to be of far-reaching philosophical significance.

Closer inspection will reveal, however, that, within a more directly 'physical' context, the dynamic conception plays another role, though partial and limited, in Bohr's 'early' view of quantum theory. In what follows, I wish accordingly to focus on this aspect of his 'early' idea of complementarity which is structurally parallel to part of the dynamic conception. Let us begin by paying attention to Bohr's occasional remarks to the effect that "the distinction between object and agency of measurement" is "inherent in our very idea of observation" (PWNB, 1:68),[8] and that this distinction stands in conflict with the quantum postulate, which implies "an interaction with the agency of observation not to be neglected." He also states, though in speaking mainly of psychological problems, that "the concepts of space

---

[7] As is the case in the previous instances, so also here my analysis differs from Meyer-Abich's in that the latter fails to distinguish between involvement (∼) and detachment (/) (see 1965, 158).

[8] This point is stressed, and referred to as "the postulate of observation," by Shingo Fujita in his analysis of Bohr's thought (1991, 70).

and time by their very nature acquire a meaning only because of the possibility of neglecting the interaction with the means of measurement" (PWNB, 1:99). This set of remarks suggest the characterization of space-time observation or coordination as a 'spectator's' detachment, and thus seem at first glance to be at odds with Bohr's overall account of complementarity in quantum theory. Was not space-time coordination characterized – in contrast to the claim of causality – as introducing "an unavoidable interference with the course of phenomena," in which the object–agency interaction cannot be neglected? Was it not therefore classified into the left column of Table 4.1, the column for the 'actor's' involvement?

This apparent discrepancy, however, offers us a clue to recognizing the following state of affairs. Bohr's remarks in the two opposite directions, taken together, may be construed as meaning that space-time coordination, which presupposes the subject–object distinction and, to that extent, is performed by a 'spectator,' turns out to have the character of being an 'actor.' In other words, space-time coordination as a detached objectification, S/O, proves to contain a moment of involvement in the phenomenon observed. This reconceptualization is carried out from a reflective point of view, S', and may thus be expressed as $S'/(S{\sim}O)'$. We can see that this state of affairs is similar, though not identical, to what I have termed the dynamic conception of complementarity concerning fields other than quantum theory, in which reflection proves, on further reflection, to have a character of involvement. Although Bohr himself does not use the term 'complementary' to refer to the above dynamic relation between space-time observation as detachment and as involvement, it is reasonable to characterize this relation as parallel to part of the dynamic conception of complementarity.

Further, this suggests a new way of understanding Bohr's 'early' use of phrases such as 'the disturbance of phenomena by observation.' As we saw in Section 2.3, from the late 1930s onward, he would regard such expressions including the term 'disturbance' as problematic and abandon their use. As also noted in Section 3.3, many commentators including Folse and Faye consider this move to be a terminological improvement, on the ground that the disturbance language misleadingly suggests that atomic objects possess inherent properties which are only posteriorly disturbed by observational interaction. I do not, however, find a sufficient reason for this widely shared assumption.[9] In my view, the term 'disturbance' – or Bohr's 'early' usage thereof – does not necessarily presuppose the existence of an inherent undisturbed state, any more than, say, the term 'division' presupposes the existence of an undivided unity or the term 'contamination' the existence of something purely uncontaminated. In what follows, I wish to show that Bohr's 'early' use of the term 'disturbance' – in a close connection with the term 'phenomenon' – may be interpreted more appropriately from the point of view of the dynamic conception of complementarity.

[9] As noted in Section 3.3, this line of critique of the disturbance language by commentators differs from Bohr's own later critique thereof.

By way of comparison, let us start by examining Bohr's usage of another term: interaction. As can be readily seen, he employs this term now in combination with 'phenomenon' – as in the clause that "any observation of atomic phenomena will involve an interaction with the agency of observation" – now in combination with 'object' – as in the phrase "a finite interaction between the object and the measuring instrument" (PWNB, 1:54, 114; cf. 93). This is also the case with his usage of the practically opposite term 'distinction,' which is now combined with 'phenomenon' as in "a sharp distinction between phenomena and their observation," now combined with 'object' as in "the distinction between object and agency of measurement" (PWNB, 1:22, 68). To this extent, the two terms 'phenomenon' and 'object' seem to be treated as synonymous, both referring to the atomic (or subatomic) system under investigation. In contrast, Bohr employs the term 'disturbance' and more or less comparable words such as 'influence' and 'interference' invariably in combination with the term 'phenomenon,' and not with 'object.' We can cite as examples the following expressions: "the idea that the phenomena concerned may be observed without disturbing them appreciably;" "[t]he unavoidable influence on atomic phenomena caused by observing them;" and "any observation necessitates an interference with the course of the phenomena" (PWNB, 1:100, 115; cf. 11). This suggests how the two terms 'object' and 'phenomenon' are at the same time distinguished and interconnected. On the one hand, the phenomenon overlaps with the object of observation, which I denote by O.[10] On the other, the term 'phenomenon' also refers to what appears as the effect of the object–agency interaction, that is, $(S{\sim}O)'$. While only the latter would survive Bohr's later redefinition of the term, these two senses of 'phenomenon' coexist in his 'early' texts. More importantly, these two meanings may be considered as interconnected through the concept of disturbance in such a way that the phenomenon in the sense of the object is 'disturbed' to *become* what appears as the effect of the observational interaction. On this interpretation, the ambiguity of the concept of the phenomenon is no misguided confusion, nor is the use of the term 'disturbance' illegitimate. What Bohr means by the disturbance of the phenomenon by observation is not a change in the object (from an undisturbed to a disturbed state) as it is viewed from a fixed standpoint of the subject. Rather, it is an alteration in the very relation between subject and object. This alteration is precisely the 'objective' side – represented by the right side of the slash – of the transition from S/O to $S'/(S{\sim}O)'$, characterized earlier as the part of the dynamic conception that is relevant to quantum theory. In other words, the term 'disturbance' refers to the circumstance that the phenomenon, initially identified with the object O by the observing subject S, becomes $(S{\sim}O)'$ in correlation with the transition of the standpoint of the subject from S to S'.[11]

[10] Held, among others, seems not to attend to this partial overlapping of the terms 'object' and 'phenomenon,' in his critique of the "lack of clarity" of Bohr's 'early' argument regarding disturbance (Held 1999, 54).

[11] This circumstance is reminiscent of Hegel's account of the dynamic structure of consciousness in the *Phänomenologie des Geistes*: He specifically argues that "as the knowledge changes, so too does the object" in such a way that "it comes to pass for consciousness that what it previously took

So far in this chapter, I have analyzed in some detail Bohr's 'early' idea of complementarity, discussing the philosophical implications of its two heterogeneous conceptions, particularly of the dynamic conception. As we have seen in the present section, the dynamic conception in this period, while largely restricted to fields outside physics, plays a partial and limited role in his interpretation of quantum theory. From the mid-1930s onward, however, Bohr would thoroughly extend the dynamic conception to the field of quantum theory, and this extension and its consequences essentially characterize his thought in the 'middle' period. This will constitute the subject of discussion in the next section.

4.3 The 'Middle' Period: A Reorganization of Complementarity

Having so far analyzed Bohr's 'early' idea of complementarity, I now proceed to the next stage, starting with his 1935 debate with EPR, and examine whether and, if so, in what way his complementarity argument underwent changes beyond the level of mere terminological refinement. As reviewed in Section 2.3, in order to demonstrate the incompleteness of quantum mechanics, Einstein, Podolsky, and Rosen introduced a thought experiment on two separate physical systems I and II. The crux of their argument is that if one knows the state of the combined system I+II, one can predict the position/momentum of system II by measuring the position/momentum of system I alone, thus "without in any way disturbing the second system." In response, Bohr admitted that there is "no question of a mechanical disturbance" of system II brought about by the measurement on system I. Yet, he stressed, "there is essentially the question of an influence on the very conditions which define the possible types of predictions regarding the future behavior of the system" (PWNB, 4:80). This introduction of a renewed notion of observational influence – as a 'non-mechanical' influence on a system that is not measured on – has been discussed abundantly by a number of commentators (e.g. Folse 1985, 150ff.; Faye 1991, 178ff.).[12]

It appears to me, however, that Bohr's above conceptual innovation is associated with a change in his argument which may be less manifest and yet is of particular significance in the present context of this study. In discussing the way in which measurement on system I influences the conditions of possible prediction about system II, Bohr deals not only with the case of position measurement, but also with the case of momentum measurement. This, along with a parallel discussion of the alternative measurement of time or energy (PWNB, 4:80f.), suggests the following change in his idea of complementarity. What Bohr means by "the finite and

---

to be the *in-itself* is not an *in-itself*, or that it was only an in-itself *for consciousness*" (1807, 72/54). This may serve as a point of departure for examining the possible conceptual linkage between Bohr's complementarity – especially its dynamic conception – and Hegelian dialectics.

[12] In Jammer's account, a crucial implication of this conceptual revision by Bohr is a "relational" and "holistic" conception of quantum mechanical states (1974, 198, 200).

uncontrollable interaction between the objects and the measuring instruments" is no longer restricted to the exchange of momentum and energy in the case of space-time observation, but covers equally the "uncontrollable displacement" in the case of momentum and energy measurement, which obliges us to renounce the control of the space-time coordination (PWNB, 4:80, 77). In this way, observational influence, reconceptualized beyond a mere mechanical disturbance, is extended not only to the case of predictions about distant systems, but also to the case in which the dynamical conservation laws are applied in general. In other words, the use of the momentum and energy conservation laws (or, in his earlier terminology, the claim of causality) is no longer simply associated with a 'spectator's' detachment, but recomprehended as having the character of an 'actor's' involvement as well. This tends to undermine the very validity of Bohr's 'early' idea of complementarity in quantum theory, marked by the static contrast between the roles of 'actor' and 'spectator.'

This conceptual change in Bohr's complementarity, occasioned by his confrontation with the EPR argument, implies the following tacit extension of his ideas. When, in the late 1930s, he points to the impossibility to "distinguish sharply between the behaviour of objects and the means of observation," or to "an absolute limit to the possibility of speaking of a behaviour of atomic objects which is independent of the means of observation" (PWNB, 2:25), he might seem simply to repeat the same language as in the 'early' period. While, however, the "limit" to the possibility of speaking of independent atomic objects or, in other words, the limit of taking the role of a 'spectator,' had earlier been restricted to the case of space-time coordination, this restriction is now removed since momentum-energy measurement is also viewed as taking the role of an 'actor.' That is, the above epistemological limit is generalized to all situations in quantum theory, including the case of the application of the dynamical conservation laws.

This is not, however, all the philosophical implications of the above alteration in Bohr's idea of complementarity. As I argued in the previous section, Bohr's 'early' account of quantum theory had implicitly contained a limited part of the dynamic conception of complementarity, namely, the notion that space-time coordination, provisionally conceived as a 'spectator's' detachment, turns out to have the character of an 'actor's' involvement. What we have seen above with regard to his response to EPR is his new view that not only space-time coordination, but also its complement, dynamical conservation, proves to have the mode of being an 'actor.' Does this not suggest that, in his transition to the 'middle' period, Bohr's dynamic line of thought comes to the foreground in his interpretation of quantum theory in such a way to cover both complementary relata? In order to examine this question, I refer to a passage quoted only partly earlier from "Discussion with Einstein," which, though written toward the end of the 'middle' period, seems to be particularly relevant in this context:

> [...] the individuality of the typical quantum effects finds its proper expression in the circumstance that any attempt of subdividing the phenomena will demand a change in the experimental arrangement introducing new possibilities of interaction between objects and measuring instruments which in principle cannot be controlled (PWNB, 2:40).

This may be rendered in more specific terms as follows: While space-time coordination involves an object–instrument interaction in the form of an "exchange of momentum and energy," any attempt to determine this exchange itself will also, as suggested above, involve an interaction which takes the form of a displacement or a "latitude in the space-time description." Further, and conversely, any attempt to determine this latter interaction by means of space-time coordination will carry with it yet another interaction, namely, another exchange of momentum and energy, thus "exclude[ing] all closer account as regards the balance of momentum and energy" (PWNB, 2:40).[13]

This state of affairs may be interpreted in terms of the 'spectator–actor' relationship as follows: Any measurement – regarding either space-time coordination or momentum-energy conservation[14] – while serving as a 'spectator's' detachment, S/O, turns out to have the character of an 'actor's' involvement, $(S{\sim}O)'$. While, for space-time measurement, this corresponds to the limited dynamic idea already contained in Bohr's 'early' account of quantum theory, the same now applies to the case of momentum-energy measurement as well. Further, this reconceptualization of detachment as a mode of involvement is carried out from a new 'spectator's' reflective viewpoint, S', corresponding to a new measurement, which in turn, on still further reflection, proves to have the character of being a new 'actor.' This leads to a series of reflections, with alternating roles of 'spectator' and 'actor,' such that one can never reach a final, unambiguous determination of the observational interactions or of the properties of the object. This series, which is indefinitely extensible though not necessarily actually extended, may for the moment be rendered as:

$$\mathrm{S/O} \rightarrow \mathrm{S}'/(\mathrm{S} \sim \mathrm{O})' \rightarrow \mathrm{S}''/(\mathrm{S}' \sim (\mathrm{S} \sim \mathrm{O})')'' \rightarrow \ldots\text{[15]} \qquad (4.2)$$

[13] See also PWNB, 2:19 and 4:87. For partly corresponding remarks from the 'late' period, see PWNB, 2:72 and 90. In the 'middle' period, Bohr no longer speaks principally – or in the same manner as earlier – of the contrast between observation and the definition of the state. He does not, as in the Como paper, associate space-time coordination uniquely with observation, and momentum-energy conservation with state definition. This is not to say, however, that he reduces definability to observability in a simply positivist manner. Rather, the contrast between observation and definition remains implicitly relevant to his thought in the following way: If one attempts to define the space-time coordinates, one must carry out an observation incompatible with the definition of momentum and energy, and vice versa.

[14] From here on, I prefer to use the phrase 'momentum-energy conservation' rather than 'dynamical conservation,' the phrase customarily used by Bohr, because the use of the latter in the present context may give rise to a confusion with my interpretive term 'dynamic' in the sense of the dynamic conception of complementarity.

[15] The third term of this expression, particularly the right side of the slash in it, is not meant to imply any physically complex state of affairs, but simply to express the circumstance that the detached reflection of the second term is reconceived as an involvement. As far as its physical meaning is concerned, the third term refers to a measurement under the same type of experimental arrangement as the first term S/O – a situation which will be made manifest below, especially by Table 4.5.

As we can readily see, this series of reflections highly resembles the dynamic series, expressed by (4.1), that I discussed with regard to fields outside quantum physics in Section 4.1.

The above series of reflections regarding quantum theory has, however, certain specific features that distinguish it from the corresponding series in other fields. First, in the above series, the alternation between the 'spectator's' detachment and the 'actor's' involvement is in a certain manner combined with an alternation between space-time coordination and momentum-energy conservation. As suggested above, if we start from a space-time measurement as detachment, this measurement turns out to have a mode of involvement, and this involvement is then to be determined by a momentum-energy measurement as detachment, which in turn proves to appear as involvement, and so on. We would proceed similarly starting from a momentum-energy measurement.

Second, we can see for the moment that while the series (4.1) for fields outside quantum physics starts with S~O, an 'actor's' involvement, the above series (4.2) starts with S/O, a 'spectator's' detachment. More importantly, however, this first term of the series, S/O, does not mean a definite or absolute beginning. Although, for the sake of convenience, I have started the series with the simple term S/O, any given measurement may serve as a reflective determination of the 'actor's' role of another measurement, which may in turn be reconceived in the same way, and so forth. For this reason, the above series of reflections is devoid not only of a definite end, but also of a definite beginning, that is, may be indefinitely extended in both directions. So as to make explicit these points, we can express the above series of reflections (4.2) by Table 4.5.

Notwithstanding the specific features just pointed out, however, the series of reflections expressed by (4.2) or Table 4.5 may reasonably be viewed as constituting a version of Bohr's dynamic conception of complementarity, a conception developed in the 'early' period mostly with regard to fields outside quantum physics. My previous comments on his dynamic conception in Sections 4.1 and 4.2 therefore basically apply to the present case as well. Specifically, Bohr's above dynamic conception regarding quantum theory conforms to neither realism nor anti-realism, nor any intermediate or eclectic position. Rather, just as in other fields of experience, it implies a suspension and radical indeterminacy between realism and anti-realism – or, better, between the realist dichotomy of subject/object and the non-realist undoing thereof. Further, in reference to Kaiser's notion of "conceptual containment" as a link between Bohr's and Kant's philosophical approaches, we can note again that, in

Table 4.5 The dynamic conception of complementarity in quantum theory

...	'Spectator' / 'Actor'	'Spectator' / 'Actor'	'Spectator' / 'Actor'	...
...	Space-time coordination	Momentum-energy conservation	Space-time coordination	...

the above dynamic conception, any description of atomic processes, while subject to a movement of conceptual containment, can never be fully contained. This is a key point at which Bohr's complementarity, including his mature account of quantum theory, diverges from Kantian transcendental philosophy.[16] We can now see how Bohr's dynamic mode of thought, initially largely limited to fields outside physics, comes to play a crucial role in his renewed interpretation of quantum theory. Bohr's transition from the 'early' to the 'middle' period is thus marked by a further radicalization of the idea of complementarity, in which his dynamic conception becomes relevant to all fields of knowledge and experience under consideration.

This extension of the dynamic conception does not, however, cover the whole character of Bohr's thought in the 'middle' period. Rather, it is no less noteworthy that the above expansion of the dynamic conception serves paradoxically to produce *another static structure* – other than the one in his 'early' complementarity argument. As we have seen, Bohr's above dynamic conception revolves around the move in which space-time coordination and energy-momentum conservation, initially conceived as of the character of being a 'spectator,' become each reconceptualized as having the mode of being an 'actor.' If, however, this double character of both complementary relata is considered not as constitutive of this dynamic movement, but, on the contrary, *in abstraction from the dynamic dimension*, it leads to the formation of a new static and onefold structure of complementarity. That is, under certain conditions, Bohr comes to juxtapose the two complementary relata as comparable items that both in part carry with them observational interactions.

This new static point of view does not, however, derive directly or straightforwardly from the dynamic conception of complementarity. Rather, it should first be examined why and in what way abstraction from the dynamic dimension takes place. In Section 2.3, we saw a series of terminological changes introduced by Bohr during the 'middle' period: the rejection of the notion of 'disturbance' and, linked with this, a redefinition of key terms such as 'phenomenon' and 'complementarity.' In my view, it is precisely these changes that serve as a conceptual setting for the formation of the new static scheme. As I indicated in the previous section, Bohr's 'early' notion of the phenomenon has the double meaning of the object O itself and what appears as the effect of the object–agency interaction, $(S{\sim}O)'$, and these two meanings may be seen as connected in such a way that the former is 'disturbed' to become the latter in correlation with the transition of the subject's standpoint from S to S'. This

[16] It seems to me that here lies a crucial difference between Bohr's and Einstein's philosophical orientations. As rightly pointed out by some commentators against common misreadings, the later Einstein is by no means a (naïve, metaphysical, or scientific) realist, but rather orients himself to a quasi-Kantian transcendental epistemology (see Murdoch 1987, 195–99; Held 1999, 219ff.; Katsumori 1992, 584ff.). While thus neither Einstein nor Bohr may be called realist, they radically differ from each other *in the way their views differ from realism*. In Einstein's view, the 'real external world' is conceptually constructed by the human mind (see Einstein 1979, 64ff./60ff.), and this construction is fully subject to what Kaiser terms conceptual containment. On the other hand, Bohr's complementarity stresses our practical involvement in world phenomena, which each time exceeds the mechanism of conceptual containment.

interpretation is essentially in line with Bohr's dynamic mode of thought, which, as we have seen above, assumes further importance in the 'middle' period. In the same period, however, he begins to make apparently contrary moves: He newly defines the term 'phenomenon' – as "the effect observed under given experimental conditions" (NBCW, 7:316) – in such a way as to separate it from the term 'object.' This move is closely associated with his rejection of phrases such as "disturbing of phenomena by observation," since the notion of disturbance had earlier served dynamically to mediate the two terms 'object' and 'phenomenon.' In this way, the atomic object and the phenomenon are now treated as distinct self-identical beings (whatever the ontological status of the object may be),[17] whose meanings are no longer ambiguous or subject to displacement. This is precisely the conceptual basis on which he redefines the very term 'complementarity' as the relation between different phenomena (corresponding to mutually exclusive experimental arrangements) concerning one and the same object. Bohr's above terminological revisions thus imply, as it were, forbidding one to say that O is disturbed to become $(S{\sim}O)'$, or that S/O goes over into $S'/(S{\sim}O)'$, or, in other words, forbidding one to speak along the lines of the dynamic conception.

Under these circumstances, there seems to be left only one way to conceive the double character of detachment (S/O) and involvement (S∼O) of either space-time or momentum-energy measurement: Each of the two complementary relata *simultaneously* has both characters of being a 'spectator' and an 'actor.' More specifically, space-time coordination serves as a 'spectator' with regard to the space-time coordinates of the object, while serving as an 'actor' with regard to the energy and momentum. In the diffraction experiment with an unmovable diaphragm, for example, one behaves as a 'spectator' in measuring the exact position of the particle, while at the same time playing the role of an 'actor' as far as the momentum of the particle is concerned. I express these two aspects as $(S/O)_q$ and $(S{\sim}O)_p$, respectively, where subscripts q and p denote the points of view – of space-time and momentum-energy, respectively – from which the subject–object relation is considered.[18] On the other hand – as illustrated by the diffraction experiment with a movable diaphragm[19] – the use of the dynamical conservation laws serves as a 'spectator' with regard to the energy and momentum, and as an 'actor' with regard to the space-time coordinates, which may similarly be written as $(S/O)_p$ and $(S{\sim}O)_q$,

[17] Needless to say, this is not to deny the fact that the term 'object' is still connected with the term 'phenomenon' through the circumstance that the object–instrument interaction forms part of the phenomenon.

[18] Here I have simply borrowed the symbols of conjugate variables to use them in a non-technical way. While q and p commonly denote position and momentum (more precisely, q a generalized coordinate and p the momentum conjugate to q), q and p here refer to the viewpoints of consideration corresponding to these physical quantities.

[19] For Bohr's account of the functions of the movable/unmovable diaphragm in the diffraction experiment, see, for example, PWNB, 4:76–78. For a philosophical analysis of this account, see Don Howard, "What makes a Classical Concept Classical? Toward a Reconstruction of Niels Bohr's Philosophy of Physics," in Faye and Folse (1994, 201–29, esp. on 214).

respectively. In this way, space-time coordination and momentum-energy conservation – more precisely, the two corresponding types of 'phenomena' in the newly defined sense – are each conceived as having both moments of detachment and involvement, depending on which set of the conjugate variables is considered.[20] The juxtaposition of these two phenomenal items, $(S/O)_q \bullet (S{\sim}O)_p$ and $(S{\sim}O)_q \bullet (S/O)_p$, constitutes Bohr's new static conception of complementarity in quantum theory, which develops in the 'middle' period and eventually becomes predominant over the dynamic conception. Since the complementary relata here stand no longer in contrast with each other, but rather appear as similar and interchangeable, I wish to designate this conception as *static-symmetrical* in distinction to the earlier static one, which may be specifically called *static-contrastive*.

We have seen so far how, in the 'middle' period, Bohr's dynamic conception of complementarity was not only extended to quantum theory, but also served – through certain conceptual mediation – to produce a new version of complementarity marked by the static-symmetrical juxtaposition of complementary relata. Not only do these two conceptions of complementarity stand in tension with each other just as the two conceptions in the 'early' period (the static-contrastive and the dynamic), but they conflict more directly with each other since they both concern the same field of quantum theory. Bohr's different conceptions of complementarity in the 'early' and 'middle' periods as considered thus far may be succinctly recapitulated by the following diagram:

The 'early' period		The 'middle' period
Static-contrastive juxtaposition		Static-symmetrical juxtaposition
		↑
Dynamic displacement	→	Dynamic displacement (extended)

This suggests a philosophically paradoxical character of the 'middle' period: Bohr's thought at this stage is marked by both a radicalization of his 'early' idea of complementarity *and* a step toward an epistemologically more classical version of complementarity.

In this section, analyzing Bohr's complementarity in terms of its conceptual structure and historical development from 1935 until around 1950, I have made several points that have not been indicated by prior studies. We have specifically seen that Bohr's complementarity underwent, not a mere terminological refinement, but

[20] As suggested by my discussion in Section 2.3, Bohr also divides the moment of the 'actor's' involvement – which he had earlier called the "influence on atomic phenomena caused by observing them" (PWNB, 1:100) – into two distinct factors: on the one hand, the experimental arrangement, which conditions the phenomenon, and, on the other, the object–instrument interaction, which takes place under that arrangement and constitutes part of the phenomenon. In this conceptual setting, as far as the phenomenon is concerned, not observation, but experimental arrangement alone plays the role of an 'actor,' while observation has the character of being an 'actor' with regard to the atomic object. Here I do not, however, further enter into this state of affairs.

a set of changes in the basic conceptual structure which consists of two conflicting but paradoxically linked moments: an extension of the dynamic conception and a formation of the static-symmetrical conception. To be sure, Bohr himself never acknowledged any significant change in his idea of complementarity and generally wrote as if his point of view were both structurally coherent and diachronically unchanged. Yet our consideration so far suggests that his thought went through a process of transformation which is much more subtle than, and yet comparable in significance to, what may be called Einstein's philosophical turn. In the next section, I will proceed to Bohr's complementarity in the 'late' period, and examine further philosophical implications of his new static conception as it would serve as the basis of his later objectivist thought.

4.4 The 'Late' Period: Toward an Objectivist Philosophy

With the conceptual reorganization of complementarity as analyzed in the previous section, Bohr did not yet arrive at the final stage of his philosophical development. As we have seen, his idea of complementarity in the 'middle' period was highly conflictual insofar as it consisted of two hardly compatible – dynamic and static-symmetrical – conceptions. From around 1950 onward, however, he began to reorient himself solely to the static-symmetrical conception, seeking on this basis to formulate a coherent philosophical approach to quantum theory and beyond.

We saw in Section 2.4 that, during the 'late' period, Bohr turns away from his earlier emphasis on the unavoidable ambiguity of language and experience, and comes to stress the unambiguous and objective character of our description of nature, specifically as provided by quantum theory. Closely associating objectivity with "unambiguous communication," he seeks to base the possibility of unambiguity on the use of "common human language" including the terminology of classical physics. On his account, by the use of such language in atomic physics, both the experimental arrangement and the phenomenon can be described unambiguously. He further maintains that "we must retain a sharp distinction between the observer and the content of the observations," which, under certain circumstances, requires "different placings" of such a distinction (PWNB, 2:91f.). This may be carried out "by including in the account of the phenomena explicit reference to the experimental conditions," which allows one to avoid "any appeal to the observing subject" and to attain a full objectivity and unambiguity (PWNB, 3:12, 7). Bohr thus claims that "[t]he notion of complementarity does in no way involve a departure from our position as detached observers of nature" (PWNB, 2:74).

I wish to start by commenting on this last remark by Bohr. This remark, characterized by the phrase "detached observers," implies that one can obtain objective knowledge of nature without necessarily being involved in it, or, in short, that we can be pure 'spectators.' Admittedly, Bohr, still in the 'late' period, repeatedly cites the dictum that "in the great drama of existence we are ourselves both actors and spectators" (PWNB, 3:15; cf. 2:63, 81). It is not that he simply abandons the distinction between the 'actor's' involvement and the 'spectator's' detachment, but rather that he deals with this distinction in a different manner than earlier. From his new concept of the phenomenon, introduced in the 'middle' period, which requires specifying the whole experimental arrangement, he draws the following implication: All moments of the 'actor's' involvement in the phenomenon must be included in the described content, or, in other words, the description of the phenomenon must be – to use Kaiser's term – 'contained.' Bohr's point is that, in this way, one can describe nature fully objectively from the standpoint of a pure 'spectator.'

This argument by Bohr may be analyzed along the lines of what I have termed the static-symmetrical conception of complementarity, developed from the 'middle' period. As we saw in the previous section, in this new conception, the two complementary relata, specifically the two types of phenomena observed under the conditions of space-time coordination and momentum-energy conservation, are juxtaposed as parallel items $(S/O)_q \bullet (S{\sim}O)_p$ and $(S{\sim}O)_q \bullet (S/O)_p$, respectively. As regards the first item, although it involves an object–instrument interaction in the form of an exchange of momentum and energy, one must and can render the description of it unambiguous by specifying the experimental arrangement under which the phenomenon is observed. One can proceed similarly with the second item as well. In so doing, one describes both complementary relata from the standpoint of a new, reflecting subject which corresponds to Bohr's own theoretical standpoint. Although Bohr in the 'late' period rarely discusses this introduction of a new subject, we can refer to his following remark from 1957, cited partly in Section 2.4:

> [...] it must be emphasized that the distinction between subject and object, necessary for unambiguous description, is retained in the way that in every communication containing a reference to ourselves we, so-to-speak, introduce a *new subject* which does not appear as part of the content of the communication (PWNB, 2:101, my emphasis).

Though originally made in the context of biological and psychological problems, this remark appears to be particularly relevant to our discussion here primarily concerning quantum theory. The "new subject" in this passage is nothing other than what I have referred to above as the reflecting subject. As objectified by this new subject, S′, the two complementary relata in quantum theory may be schematically expressed as

$$S'/[(S/O)_q \bullet (S \sim O)_p V (S \sim O)_q \bullet (S/O)_p]. \tag{4.3}$$

That is to say, the subject S′ is faced in a detached manner with the disjunction of the two complementary relata. We can thus see that Bohr's objectivism in the 'late' period may in no way be characterized as a simple elimination of subjective elements. Rather, it proceeds with a reorganization of subject–object relations, including the introduction of a new subject S′, in such a way as to ensure this subject's privileged reflective standpoint.

What is crucial here is that, in Bohr's view, the state of affairs expressed above by (4.3) will *not* lead to a series of further reflections in which the detached reflection by S′ would turn out to be another mode of involvement, and so forth. In other words, the expression (4.3) is conceived as complete and closed on itself, and not as the initial phase of an indefinitely extensible series in which the description of the phenomena would each time be altered and displaced. This supposed nonoccurrence of a dynamic series is to be understood in two senses: First, within the physical context of quantum theory, the series (4.2) in the previous section, involving the alternation between space-time coordination and momentum-energy conservation, is no longer supposed to occur in its dynamic sense. Second, nor are epistemological reflections *on* quantum theory supposed to form a dynamic series as suggested in Section 4.2. Bohr's basic reasoning for this may be rendered as follows: Complementary description, which reflectively incorporates in itself all moments of being an 'actor,' comes to serve as a pure 'spectator' free of any character of being an 'actor.' The reflective, theoretical subject S′, embodying this pure 'spectator,' objectifies the complementary phenomena once and for all, without giving rise to any further series of reflections. Bohr's complementarity thus no longer implies a circumstance comparable to the "licentiate's" interminably entangled 'I's' in Møller's novel, but rather proceeds, like the "philistine," along the "broad highway" of the objective description of nature.

Our consideration so far indicates that Bohr's thought in the 'late' period is characterized by the fact that the dynamic conception of complementarity entirely gives way to the static-symmetrical conception, which serves as the basis of a new objectivist philosophy. In this way, his final idea of complementarity turns away from his earlier radical philosophical orientation, shifting to a more traditional line of thought. This change is not, however, a simple conservative regression, but rather, as suggested above, a paradoxical process through which his deepened insight into the moments of being 'actors' turns to an apparent overcoming of these moments, thus serving to restore the privileged standpoint of a pure 'spectator.'[21] In the rest of this section, I wish to make additional comments on this 'late'-period idea of complementarity with reference to some of the prior interpretations of Bohr's thought that were addressed in Chapter 3 and again earlier in this chapter.

(1) In Section 4.2, speaking of Bohr's 'early' idea of complementarity, I pointed to a partial validity of von Weizsäcker's distinction between parallel and circular complementarity, and, in particular, similarities as well as differences between his notion of circular complementarity and what I term the dynamic conception of complementarity. These points still largely hold for Bohr's 'middle'-period version of complementarity, which, in my interpretation, not only retains but also extends his dynamic mode of thought. As noted in Section 3.1, however, von Weizsäcker's account – specifically his concept of circular complementarity – was, in the 1950s, flatly rejected by Bohr. To be sure, von Weizsäcker's interpretation

[21] In Katsumori (1992), I have discussed a conceptual mechanism responsible for Einstein's philosophical turn that, though in a different physical context, bears some similarities to Bohr's case considered here. In particular, while Einstein's special theory of relativity revolves around the process through which the observers' experiences are intersubjectively structured, the general theory has recourse to a mathematical guarantee (in the form of tensor calculus) of the structure of intersubjectivity, which eventually leads to the notion that concept formation is carried out, as it were, by a higher-level knower transcending the observational context (see Katsumori 1992, esp. 589).

of complementarity in quantum theory contains the misunderstanding (taken over from Heisenberg) of Bohr's concept of 'the claim of causality' as a description in terms of the wave function. Yet Bohr's rejection may not solely or primarily be due to *this* misconception on von Weizsäcker's part. We cannot ignore the fact that Bohr's negative reaction took place in the 1950s, when, as suggested above, his whole idea of complementarity was in effect reduced to the static-symmetrical conception, a kind of 'parallel' juxtaposition of phenomena. It is no wonder that Bohr definitely disowned the circular – or, more generally, any non-parallel – notion of complementarity insofar as he judged it from his 'late'-period viewpoint. The question remains open, however, whether and to what extent he would have accepted von Weizsäcker's – or, for that matter, my – interpretation if it had been presented earlier, when he still retained the dynamic conception of complementarity.

(2) We have seen in foregoing sections that Bohr's earlier conceptions of complementarity, namely the static-contrastive and the dynamic conceptions, subscribe to neither realism nor anti-realism, but rather reconceive the very relation between them – or, more precisely, between the realist separation of subject/object and the non-realist undoing thereof – in the framework of complementarity. This no longer applies, however, to the static-symmetrical conception, on which his philosophical thought in the 'late' period is based. For this conception, which revolves around neither the contrast nor the alternation between the roles of 'spectators' and 'actors,' but centers on the juxtaposition of two comparable relata, can no longer be viewed as concerned with the relation of conflict between the subject/object dichotomy and its disruption.

From this there arises the possibility that Bohr's complementarity in the 'late' period simply conforms to a particular philosophical position such as realism or anti-realism. This issue will, however, depend to a large extent on how we define realism and other philosophical positions. On the one hand, if by realism we understand a fixed and unambiguous separation between the theoretical subject and *its* object – S' and $[(S/O)_q \bullet (S{\sim}O)_p \vee (S{\sim}O)_q \bullet (S/O)_p]$ in expression (4.3) – then Bohr's 'late'-period thought may be appropriately characterized as a form of realism. If, on the other, as in most other commentators' discussions and debates, we focus on the question of the reality of the atomic object O, then his orientation can still hardly be unambiguously identified with any such philosophical position. To be sure, Henry Folse, for example, is correct in pointing out that, in Bohr's later account, complementary phenomena or complementary kinds of information are obviously "about the same object" (Folse 1985, 237). From this we cannot immediately conclude with Folse, however, that this object must be an "independent reality" which "causes" the complementary phenomena (1985, 238f.). In other words, Bohr's strongly objectivist tendency in the 'late' period does not necessarily imply a realist position regarding the existence of the atomic object. The ontological status of the object appears to me to remain largely underdetermined in Bohr's account of complementarity.[22] As far as its quantum-mechanical description is concerned, the object O cannot be observed as such, but serves as something to which the two complementary types of phenomena refer in common. To be sure, if this object is viewed as an entity lying behind, and/or causally giving rise to, the phenomena, it might be called an independent reality in Folse's sense. If, however, the object is considered strictly inaccessible to sense experience, it could rather be likened to the Kantian "thing-in-itself" or "noumenon." This similarity might then serve as a clue for exploring further possible intersections between Bohr's 'late'-period idea of complementarity and Kantian philosophy. In yet another interpretation, however, one could altogether deprive the object of independent existence and regard it as a pure construct of thought, in which case the subject/object separation (S/O) and their interaction (S∼O) could also be viewed as conceptual constructs. This reading of the later Bohr as oriented toward phenomenalism or positivism, offered by commentators including Beller and Held, is indeed possible, but not necessarily plausible. The question can thus hardly be settled which of these (realist, Kantian, and phenomenalist) and other possible interpretations does the most justice to Bohr's complementarity in the 'late' period.

(3) It appears to me, however, that more crucial to the interpretation of Bohr's thought in this final period is the notion of "conceptual containment" introduced by Kaiser. As we saw in foregoing sections, what Kaiser terms conceptual containment – the mechanism of "including within the concept itself the conditions and limitations under which the concept is made" – indeed has some relevance to, yet fails to characterize adequately, Bohr's complementarity in the 'early' and 'middle' periods. His dynamic conception of complementarity, in particular, is such that any description of phenomena, while subject to a movement of conceptual containment, can never be fully contained.

In the 'late' period, however, as the dynamic conception recedes and disappears, Bohr comes to hold that full containment is both necessary and attainable. His final idea of complementarity as expressed by (4.3) implies the claim that the reflecting subject S' serves as a pure 'spectator' that 'contains' the description of the complementarity phenomena once and for all by including in it all moments of the 'actor's' involvement. It is only in this stage that Kaiser's interpretation of complementarity in terms of conceptual containment largely accords with Bohr's own thought, and that, to this extent, his analogy with Kant's philosophy, based on the notion of containment, also holds. This is not meant to suggest, however, that Bohr comes close to Kant in his overall philosophical outlook, including the view on the ontological status of the atomic object as discussed in the above subsection (2). While many commentators have centrally concerned themselves with this ontological question regarding the object and, on this basis, sought to identify Bohr's position with realism, anti-realism, or some other standpoint, it appears to me that the above question

[22] In a 1954 speech, Bohr remarks that "conceptions like realism and idealism find no place in objective description as we have defined it" (PWNB, 2:79).

of whether and how his viewpoint of complementarity allows for full conceptual containment is more essential for the understanding of his thought and its historical changes. In other words, arguably the most important alternative in this interpretive context is not so much between 'realism and anti-realism,' but rather between 'the possibility and the impossibility of full containment.' Once the notion of full containment – correlative with the idea of a pure 'spectator' – is established, the conventional opposition between realism and anti-realism seems to be at most of secondary significance.

In this chapter, I have offered, from my own interpretive point of view, an analysis of Bohr's idea of complementarity in both its structural and its historical dimensions. I have shown, in particular, how his philosophical path during his career may be grasped and characterized appropriately in terms of three distinct – static-contrastive, dynamic, and static-symmetrical – conceptions of complementarity and their mutual tensions and vicissitudes. With these results of my analysis here in mind, I will seek, in the remaining part of the book, to situate Bohr's complementarity in a broad recent and contemporary philosophical context by examining its possible conceptual intersections with hermeneutics and then with deconstruction.

Chapter 5
Intersections with Hermeneutic Philosophy

In Section 3.4, we reviewed a number of prior interpretations of Bohr's complementarity that concern themselves with its relation to the thought of various modern philosophers. These philosophers were mostly anterior to Bohr's times, although I also briefly touched on the issue of his possible links with contemporaneous logical positivism. Some other studies on Bohr, not mentioned earlier, indeed point to the relevance of still other twentieth-century philosophical approaches, and yet few of them seem to me to have particular bearing on the thematic of the present study. Edward M. MacKinnon, for example, argues that Bohr anticipated "some of the key features [of philosophy] later developed in Wittgenstein's *Philosophical Investigations*," with its major thesis that "the meaning of a word is its use in the language" (MacKinnon 1985, 115; Wittgenstein 1958, §43).[1] In my view, however, this account overlooks an important difference between the two thinkers' ideas, namely the fact that, rather than reducing the analysis of a word to its use, Bohr sets the two – use and analysis – precisely in a complementary relationship in the sense of mutual exclusion and joint completion. Generally speaking, except for a few studies including Plotnitsky's to be discussed in Chapter 6, prior research on the relation between Bohr's thought and recent or contemporary philosophy has apparently failed to yield remarkable outcomes.[2]

In this chapter, I will take up a philosophical orientation known as hermeneutics, and reexamine Bohr's idea of complementarity in its possible intersections with post-Heideggerian hermeneutic philosophy. Needless to say, hermeneutics has historically developed in close and almost exclusive connection with the fields of the humanities. Among hermeneutic philosophers, there was a strong tendency – in a way correlative with the seemingly opposite tendency of scientism – to view the natural and in particular the physical sciences as simply objectifying and explanatory forms of knowledge, thus 'relegating' them to the periphery of their philosophical

---

[1] MacKinnon himself rather inaccurately renders this idea of Wittgenstein: "The meaning of a word is determined by its usage in language" (1985, 115). Here I do not, however, enter into the difference between 'use' and 'usage' or any other interpretive issues on Wittgenstein's philosophy. See also MacKinnon (1982, esp. 357–60).

[2] Paul A. Komesaroff, in particular, has attempted to connect Bohr's complementarity with structuralism, though apparently without much success (see 1986, 267ff.).

M. Katsumori, *Niels Bohr's Complementarity*, Boston Studies in the Philosophy of Science 286, DOI 10.1007/978-94-007-1748-0_5,

thematic. Admittedly, in recent decades, this tendency has been questioned and challenged by new hermeneutic approaches that take seriously the possible hermeneutic dimension of natural science.[3] Nevertheless, on the whole, hermeneutics still mostly focuses on the human and historical studies, and its relevance to physical science is yet to be widely recognized. Under these circumstances, I wish to approach Bohr's thought below so as to bring about a new moment of intellectual exchange between the apparently largely disparate fields.[4] In Chapter 4, I analyzed Bohr's complementarity with special focus on the relation between the roles of 'spectators' and 'actors,' or, to use my own terms, between detachment and involvement. This notion of complementarity as an 'actor–spectator' relationship is reminiscent of the relation between "belonging" and "alienation" or "distanciation" – a central concern of hermeneutic philosophers such as Hans-Georg Gadamer and Paul Ricoeur. In the following first and second sections, I accordingly outline the basic ideas of Gadamer and Ricoeur, respectively, whose work is commonly recognized as representative of post-Heideggerian hermeneutic philosophy. In Section 5.3, I examine the possible conceptual intersections of Bohr's complementarity with these hermeneutic approaches. Toward the end of the chapter, I also briefly comment on Karl-Otto Apel's and Patrick Heelan's philosophy in relation to Bohr's thought.

5.1 Gadamer's Philosophical Hermeneutics

Etymologically deriving from the Greek verb *hermeneuein*, the term 'hermeneutics' earlier referred to the science and art of interpreting texts, notably biblical scripture and other classical texts (see Mueller-Vollmer 1985, 2). Hermeneutics was accorded an auxiliary status in relation to theology, jurisprudence, and the other scholarly fields it served. With the work of nineteenth-century philosophers such as Schleiermacher and Dilthey, however, it was expanded and transformed into a general theory of "understanding," although the interpretation of texts remained its paradigmatic focus (see Mueller-Vollmer 1985, 8ff.). Specifically, Wilhelm Dilthey's hermeneutics was designed to serve as a methodological foundation of the human studies, which he characterized as oriented to the understanding of inner human life in contradistinction to the natural sciences aimed at "explaining" outer objects. Further, in the twentieth century, hermeneutics assumed a still more basic philosophical significance: Martin Heidegger reconceived understanding as constitutive of the existence of the human being or "*Dasein*," thereby making hermeneutics a methodological cornerstone of his "fundamental ontology" (see 1927, 12/32, 37/62). That is, in order to approach the meaning of Being (*Sein*), he set out to analyze, through interpretation, the prior understanding that *Dasein* possesses with regard to Being, the Being of itself and the world in which it finds itself (see Mueller-Vollmer 1985, 33). This ontological dimension of hermeneutics

---

[3] A number of such studies are included in Fehér, Kiss, and Ropolyi (1999).

[4] Brock (2003), mentioned in Section 3.4 (note 54), points briefly to some conceptual "similarities" between Bohr's complementarity and hermeneutic philosophy as represented by Gadamer's work, and yet refrains from further attempting to connect the two (2003, 265).

introduced by Heidegger has formed a crucial background for a further development of hermeneutic philosophy.

Under Heidegger's profound influence, Hans-Georg Gadamer (1900–2002) developed his project of "philosophical hermeneutics," which sought, as it were, to combine the Heideggerian ontological motif and the more traditional hermeneutic concern with text interpretation (see Ricoeur 1986, 97/71, 342/277). In this section, I outline Gadamer's hermeneutics as presented in his most systematic work of 1960, *Wahrheit und Methode* [*Truth and Method*], and other related texts, with special focus on the concepts of "belonging" and "alienation."

Following the Heideggerian view of understanding as "the original form of the realization of *Dasein*" (Gadamer 1972, 245/259), Gadamer pursues the thematic of understanding (*Verstehen*) with regard to three spheres of experience: aesthetics, history, and language. In what follows, I will touch only briefly on the aesthetic sphere before proceeding to the other two. Discussing the experience of art, Gadamer develops a critique of what he calls "aesthetic consciousness," which has its roots in the modern Enlightenment. Modern aesthetic consciousness, he argues, regards art merely as a form of perceptual enjoyment separate from "the immediate truth-claim that proceeds from the work of art itself" (1967ff., 1:102; trans. 1976, 5).[5] As opposed to this subjective form of aesthetics, he recomprehends the experience of art as the understanding of a world, which consists not in the subject's grasping of something objective, but in our belonging to the truth disclosed by the work of art. This is highlighted particularly by the play (*Spiel*) structure of art. In Gadamer's account, play is not a subjective attitude or state of mind, but "the mode of being of the work of art itself," in which the dichotomy of subject and object is no longer maintained (1972, 97/101). In the theatrical play, for example, the spectator takes "the place of the player" and "the difference between the player and the spectator is [...] suspended" (1972, 105/110). That is, despite the "distance" between the play and himself, the spectator "*belongs* to play" (1972, 110/116, my emphasis; cf. 123/130) or is "involved" therein (1972, 118/125; cf. 124/132f.).

This idea of "belonging" is crucial throughout Gadamer's hermeneutics, specifically in his analysis of the experience of history. In order to show how the Heideggerian ontological insight into understanding applies to "the understanding of historical tradition" as well (1972, 250/264), Gadamer starts with an endeavor to rehabilitate the concept of "prejudice (*Vorurteil*)." In Gadamer's view, the current "negative connotation" of the word 'prejudice' goes back, again, to the modern Enlightenment (1972, 255/270). The Enlightenment is indeed characterized by its attempt to overcome all prejudices, which it defines as judgments that are not grounded on reason. Linked with this rejection of prejudice, the Enlightenment regards "tradition" as antithetical to rational consciousness.[6] According to Gadamer,

---

[5] According to Gadamer, aesthetic consciousness, along with historical consciousness, represents an "alienation when compared to the authentic experience that confronts us in the form of art itself" (1967ff., 1:102; trans. 1976, 5).

[6] In Gadamer's account, this is also connected with the Enlightenment's negative view of "authority" as "a source of prejudices" (1972, 256/271; cf. 1976, 9).

this discrediting of prejudice and tradition by the Enlightenment has for a long time seriously affected the historical sciences.[7]

Gadamer contends that the Enlightenment's "global demand" of overcoming all prejudices is itself a prejudice (1972, 260/276). For, in rejecting all prejudices, the Enlightenment not only overlooks the fact that prejudices are "not necessarily [. . .] erroneous," but crucially misses the fact that all human understanding involves prejudice as prejudgment, and that without prejudice there would be no understanding. Prejudice indeed "constitute[s] the initial directedness of our whole ability to experience" (1967ff., 1:106; trans. 1976, 9). The Enlightenment's "prejudice against prejudices in general" (1972, 255/270, trans. mod.; cf. 260/276) is closely bound up with its misguided antithesis between tradition and reason. According to Gadamer, tradition is not opposed to reason, but rather serves as "an element of freedom and of history itself," since preservation is "an act of reason" that is "as much freely chosen as are revolution and renewal" (1972, 265f./281f. cf. 261/277).[8] Correlatively, reason itself "exists for us only in concrete, historical terms – i.e., it is not its own master but remains constantly dependent on the given circumstances in which it operates" (1972, 260/276).

This suggests how, for Gadamer, the notion of our "belonging" to the historical world bears special significance. In his account,

> [. . .] our usual relationship to the past is not characterized by distancing and freeing ourselves from tradition. Rather, we are always situated within traditions, and this is no objectifying process – i.e., we do not conceive of what tradition says as something other, something alien (Gadamer 1972, 266/282; cf. 285/301f.).[9]

In short, "history does not belong to us; we belong to it" (1972, 261/276; cf. 274/290). This "belonging (*Zugehörigkeit*) to tradition," closely linked to

[7] By contrast, in Gadamer's account, Romanticism rightly opposed the Enlightenment's rejection of prejudice and tradition. He criticizes Romanticism, however, insofar as it shares with the Enlightenment "the fundamental schema of the philosophy of history," namely "the schema of the conquest of mythos by logos" (1972, 257/273). In an "abstract opposition to the principle of enlightenment," Romanticism "conceives of tradition as an antithesis to the freedom of reason and regards it as something historically given, like nature" (1972, 265/281). In other words, Romanticism "shares the presupposition of the Enlightenment and only reverses its values" (1972, 258/273).

[8] In a related context, Gadamer argues also against the Enlightenment's "antithesis between authority and reason" (1972, 261/277). In his account, authority has "nothing to do with blind obedience to commands," but instead "rests on acknowledgment and hence on an act of reason itself" (1972, 264/279).

[9] To put it differently, "we cannot extricate ourselves from the historical process, so distance ourselves from it that the past becomes an object for us." Rather, "[w]e are always situated in history" (1967ff., 1:158; see Ricoeur 1986, 346/281). In Gadamer's account, this is epitomized by the case of what is called "classical": "The classical is something that resists historical criticism because its historical dominion, the binding power of the validity that is preserved and handed down, precedes all historical reflection and continues in it" (1972, 271/287).

Heidegger's notion of "being-in-the-world,"[10] proves to constitute a "prior condition" of the understanding of the past (1972, 312/328, 275/291; cf. 248/262). For, Gadamer maintains, understanding is possible insofar as the act of understanding is not isolated from, but already connected with, that which is understood, so that even misunderstanding presupposes a "deep common accord" (1967ff., 1:104; trans. 1976, 7). This is the reason why "belonging to a tradition is a condition of hermeneutics" (1972, 275/291).[11] From this it also follows that "the prejudices of the individual, far more than his judgments, constitute the historical reality of his being" (1972, 261/276f.; cf. 1976, 9). For "the meaning of 'belonging' – i.e., the element of tradition in our historical-hermeneutical activity – is fulfilled in the commonality of fundamental, enabling prejudices" (1972, 279/295).

Gadamer's emphasis on belonging does not mean, however, that our relation to a tradition is a simple unity with the past. In his view, while understanding has a bond to the subject matter mediated by tradition, "hermeneutical consciousness is aware that its bond to this subject matter (*Sache*) does not consist in some self-evident, unquestioned unanimity" (1972, 279/295). Rather, the tradition has a moment of "strangeness" owing to our temporal and "historical distance" from the past, which is "characteristic of our hermeneutic situation" (1972, 284/300).[12] It follows, continues Gadamer, that "[h]ermeneutic work is based on a polarity of familiarity and strangeness (*Fremdheit*)," on the tension between belonging to a tradition and "being a historically intended, distantiated object." Indeed, "[t]he true locus of hermeneutics is this in-between" (1972, 279f./295). This is not to say, however, that, in Gadamer's view, belonging and distance, or the familiar and the strange, stand on an ontologically equal footing. Rather, "[m]isunderstanding and strangeness are not the first factors," but "[o]nly the support of familiar and common understanding makes possible the venture into the alien" (1967ff., 1:111; trans. 1976, 15). In this sense, Gadamer does prioritize the familiar over the strange, and belonging over distance.

---

[10] Rejecting the modern dichotomy of subject and object, Heidegger characterizes *Dasein*'s fundamental mode of being as "being-in-the-world (*In-der-Welt-Sein*)," which signifies the "unitary phenomenon" of one's inhabiting the world (1927, 53/78). As he puts it, "[i]t is not the case that man 'is' and then has, by way of an extra, a relationship-of-Being towards the 'world,'" but rather that taking up such a relationship "is possible only *because* Dasein, as Being-in-the-world, is as it is" (1927, 57/84). See also Ricoeur (1986, 46/30).

[11] Gadamer characterizes this belonging to a tradition even as the "natural relation to the past" (1972, 266/282).

[12] In the earlier editions of *Wahrheit und Methode*, Gadamer remarks that "[i]t is only temporal distance that can solve the question of critique in hermeneutics, namely how to distinguish the *true* prejudices [. . .] from the *false* ones" (1972, 282). As he himself comments in a later (the 5th) edition, however, Gadamer has "softened" this passage into: "Often temporal distance can solve [. . .]." The reason he gives for this change is that "it is distance, not only temporal distance, that makes this hermeneutic problem solvable" (1986ff., 1:304). See the 1989 edition of the English translation, based on this fifth German edition, p. 298. As is suggested here, Gadamer extends the notion of distance from temporal and historical distance to all types of interpretive distance that serve the same function (see 1987, 125). See Johnson (2000, 30).

Further, in Gadamer's view, it would also be misguided to regard tradition simply as a "permanent precondition" under which we are already situated. Rather, "we produce [tradition] ourselves inasmuch as we understand, participate in the evolution of tradition, and hence further determine it ourselves" (1972, 277/293; cf. 274f./290, 288/304). In thus "participating (*Einrücken*) in an event of tradition" (1972, 274f./290), the understanding of the past is itself a "historically effected process" (1972, 283/300, trans. mod.). This leads to Gadamer's concept of "effective-history (*Wirkungsgeschichte*)," which implies that we understand history, not as a subject separated from the object, but as a consciousness that is itself "historically effected," thus illuminating our "hermeneutical situation" from within that situation (1972, 285/301).[13] As regards historical research, in particular, "the abstract antithesis between tradition and historical research, between history and the knowledge of it" cannot be upheld. Rather, "[t]he effect (*Wirkung*) of a living tradition and the effect of historical study must constitute a unity of effect" (1972, 267/282).[14]

This notion of effective-history allows us to understand better how Gadamer views the problem of the "hermeneutic circle." In his account, nineteenth-century hermeneutics discussed the circular structure of understanding "always within the framework of a formal relation between part and whole." In other words, it rightly but simply pointed out that "the movement of understanding is constantly from the whole to the part and back to the whole." It is Heidegger who, going beyond this "methodological" notion of the hermeneutic circle, introduced a new, "ontological" conception thereof. That is, Heidegger recomprehended the circle as the circumstance that every interpretation draws on anticipations of understanding, which is constitutive of *Dasein*'s mode of being.[15] Following this Heideggerian move in the context of historical experience, Gadamer conceives the circle as the relation between text interpretation and "the anticipatory movement of fore-understanding" which is grounded in belonging to a tradition.[16] As he puts it:

[13] Linked with this idea of effective-history, Gadamer discusses the concept of "application," maintaining that "understanding always involves something like applying the text to be understood to the interpreter's present situation" (1972, 291/308; cf. 381/403). In "legal hermeneutics" (1972, 307/324), for example, application lies in "concretiz[ing] the law in each specific case" (1972, 312/329). Further, for this reason, "the text, whether law or gospel, if it is to be understood properly [...] must be understood at every moment, in every concrete situation, in a new and different way" (1972, 292/309).

[14] In this sense, Gadamer remarks, "historical research is not only research, but the handing down of tradition" (1972, 268/284).

[15] In Heidegger's account, this "fore-structure" of understanding consists of three moments: "fore-having (*Vorhabe*)," "fore-sight (*Vorsight*)," and "fore-conception (*Vorgriff*)" (1927, 150/191, 152/194).

[16] Jean Grondin, however, points to differences between Heidegger's and Gadamer's views of the hermeneutic circle: One of the major differences is that while Heidegger avoids speaking of the circle of whole and part, Gadamer associates the idea of circularity with the idea of the *coherence* of whole and part, which he considers to be a "criterion of correct understanding" (Gadamer 1972, 275/291). Jean Grondin, "Gadamer's Basic Understanding of Understanding," in Dostal (2002,

> The circle, then, is not formal in nature. It is neither subjective nor objective, but describes understanding as the interplay (*Ineinanderspiel*) of the movement of tradition and the movement of the interpreter (Gadamer 1972, 277/293).

This interplay, in Gadamer's account, offers the reason why "understanding is not merely a reproductive but always a productive activity as well," or, in other words, why "we understand in a *different* way, *if we understand at all*" (1972, 280/296f.).

Closely associated with the idea of effective-history is also the thematic of the "horizon" and "the fusion of horizons." According to Gadamer, in understanding in general, we must place ourselves in the horizon of what is to be understood. Historical understanding therefore involves "transpose[ing] ourselves into the historical horizon" (1972, 286/303). This does not mean, however, that historical consciousness simply "pass[es] into alien worlds unconnected in any way with our own" (1972, 288/304). Rather, our encounter with the past, in which we are faced with its "historical alterity," allows us to "test all our prejudices" and thereby to modify and enlarge our own horizon (1972, 286/302, 289/306). In this way, our horizon or "the horizon of the present is continually in the process of being formed" in conjunction with the horizon of the past, thus "rising to a higher universality that overcomes not only our own particularity but also that of the other" (1972, 288/305f.). Gadamer designates this process of joining or merging together of horizons as a "fusion of horizons (*Horizontverschmelzung*)" (1972, 289f./306)."[17] Historical understanding is thus characterized by the fusion of the horizons of the present and the past. In Gadamer's account, "[t]o bring about this fusion in a regulated way is the task of [...] historically effected consciousness" (1972, 290/307).

Having so far discussed understanding in the experience of history, Gadamer now proceeds to the question of the *medium* of understanding, namely, "language." He sets up a general thesis: "Being that can be understood is language" (1972, 450/474; cf. 1981, 25).[18] Language is not a tool or a mere means of communication, but a medium, and indeed "the universal medium in which understanding occurs" (1972, 366/389; cf. 1976, 15). As regards historical understanding, in particular, "the essence of tradition is to exist in the medium of language, so that

---

36–51, esp. 47–50). Georg W. Bertram offers yet another reading of Gadamer's conception of the circle and its differences from Heidegger's (see Bertram 2002, 51–56).

[17] Gadamer also maintains that the two horizons are not truly "closed horizons," but that "[e]verything contained in historical consciousness is in fact embraced by a single historical horizon" (1972, 288/304). As he comments himself, this might seem to be at odds with the idea of the fusion of horizons itself: If everything were already within a single horizon, no fusion would occur any longer. Gadamer appears to hold, however, that the two horizons are never *completely* separate, and yet that they may be fused insofar as the horizon of the past is experienced as other by the interpreter (1972, 290/306). It is also worth noting that, in Gadamer's later account, the above "single horizon" is not "an abiding and identifiable 'one'" (1987, 119).

[18] Gadamer renders this point by saying that "things bring themselves to expression in language" (1967ff., 1:69; trans. 1976, 81). See also 1972, 394/417.

the preferred *object* of interpretation is a verbal one" (1972, 367/389). Not only, however, does the object of understanding have its life within language, but also the subject who understands is situated within language. For this reason, "[v]erbal interpretation is the form of all interpretation, even when what is to be interpreted is not linguistic in nature – i.e. is not a text but a statue or a musical composition."[19] For the forms of interpretation in such cases, while not verbal, are "in fact presuppose language (*Sprachlichkeit*)" (1972, 376/398). In this sense, language is "universal" and "all-encompassing" (1967ff., 1:99; trans. 1976, 67). As Gadamer puts it,

> All understanding (*Verstehen*) is interpretation (*Auslegen*), and all interpretation takes place in the medium of language that allows the object to come into words [. . .] (Gadamer 1972, 366/389).[20]

Thus, language is "the fundamental mode of operation of our being-in-the-world and the all-embracing form of the constitution of the world" (1967ff., 1:101; trans. 1976, 3).[21]

By language, Gadamer essentially means spoken language.[22] In his account, language "lives in speech," and is basically of a "conversational character" (1972, 382/404; 1987, 122; cf. 124).[23] Not all understanding, however, takes place in immediately conversational situations. Rather, in many cases, understanding, and particularly understanding of the past, is mediated by "writing (*Schrift*)." In Gadamer's view, it is with regard to writing that the problem of distance or alienation becomes most manifest,[24] because writing is characterized by its "detachment both from the writer or author and from a specifically addressed recipient or reader" (1972, 369/392). As "the abstract ideality of language" (1972, 370/392), writing is "a kind of alienated speech" (1972, 371/393; cf. 368/390). In contrast to speech as living language, writing is, as it were, "the dead trace of meaning" (1972, 156/164). For this reason,

[19] Gadamer at times, however, uses the term 'text' in a wider sense including "a picture, an architectural work, even a natural event" (1985, 111).

[20] As Mueller-Vollmer points out, the relation between the terms 'understanding' and 'interpretation' in Gadamer's work is not unambiguous (Mueller-Vollmer 1985, 40). On the one hand, Gadamer tends to equate the two by saying that "understanding and interpretation are, in the final analysis, one and the same" (1972, 366/388, trans. mod.; cf. 380/403). On the other, he at times partially differentiates between them, maintaining that "interpretation is the explicit form of understanding" (1972, 291/307; cf. 376/398).

[21] It also follows that "the fusion of horizons that takes place in understanding is actually the achievement of language" (Gadamer 1972, 359/378).

[22] Gadamer even excludes writing from the very concept of language. This is illustrated by such remarks as: "the text can be transformed back into language (*Sprache*)" (1972, 368/391); "in relation to language (*Sprachlichkeit*), writing seems a secondary phenomenon" (1972, 370/392; cf. 382/404).

[23] Gadamer remarks most briefly that "language is conversation" (1985, 106; cf. 113).

[24] Admittedly, in Gadamer's view, "[i]t is not only the written tradition that is estranged," but "everything that is no longer immediately situated in a world" (1972, 157/165).

> [. . .] its signs need to be transformed back into speech and meaning. Because the meaning has undergone a kind of self-alienation through being written down, this transformation back is the real hermeneutical task (Gadamer 1972, 371/393).[25]

In other words, understanding of the past consists in bringing back the written tradition "out of the alienation (*Entfremdung*) in which it finds itself and into the living present of conversation" (1972, 350/368).[26] In this way, "[t]he sign language of writing refers back to the actual language of speech" (1972, 370/392, trans. mod.). As we can see, this relation between speech and writing exemplifies Gadamer's general conception of the relation between belonging and alienation.

In this section, we have traced the main lines of argument of Gadamer's philosophical hermeneutics with special attention to the conceptual opposition of belonging/alienation. To be sure, despite his rejection of the Diltheyian dichotomy between understanding in the humanities and explanation in the natural sciences, Gadamer follows the tradition of hermeneutics in focusing on the human and historical sciences as paradigmatic fields for the thematic of understanding. While maintaining with Heidegger that "the natural scientific mode of knowledge appears [. . .] as a subspecies of understanding" (Gadamer 1972, 245/259; cf. xxviiif./xxif., 277/333),[27] natural science remains outside his major scope of philosophical interest. Yet this does not mean that Gadamer's hermeneutics has little relevance to philosophical problems concerning natural science. Later in this chapter, I will suggest such relevance in my analysis of the conceptual intersections between Gadamer's hermeneutics and Bohr's complementarity. Prior to this, however, let us in the next section turn to the work of another major hermeneutic philosopher: Paul Ricoeur.

5.2 Ricoeur's Hermeneutics of the Text

The work of Paul Ricoeur (1913–2005) constitutes another landmark in the post-Heideggerian development of hermeneutic philosophy. Although his work successively covers a wide range of themes,[28] I wish here to restrict myself

---

[25] For parallel passages, see Gadamer (1972, 156/163, 372/394, 449/474; 1981, 34, 41ff.).

[26] According to Gadamer, reading means essentially "reading aloud," and even "silent reading" involves "an inner speaking" (1972, 153/160).

[27] In his article "The Universality of the Hermeneutical Problem," Gadamer states that the hermeneutical dimension is not restricted to the humanities, but "encompasses the entire procedure of science" (1967ff., 1:107; trans. 1976, 11). The early Heidegger rather briefly discusses science, specifically "mathematical natural science," in 1927, 356–64/408–15, esp. p. 362/413f. According to Joseph Rouse, this account by Heidegger still rests on "the traditional theory-dominant understanding of science," which runs counter to his own overall practice-oriented hermeneutic philosophy (1987, 72–97, esp. 79).

[28] During his career, Ricoeur passed from the pre-hermeneutical phenomenology of the will to the hermeneutics of the symbol and then of the text, and further to hermeneutic investigations of human action, history, narrative, and other categories (see Ricoeur 1995, 38, 43). It should be noted that,

mostly to his hermeneutics of discourse and the text, developed largely during the 1970s, which may be viewed as one of the cornerstones of his overall hermeneutic thought. I mainly draw on his book *Interpretation Theory: Discourse and the Surplus of Meaning* (1976), several articles included in *Du texte à l'action: Essais d'herméneutique, II* [*From Text to Action: Essays in Hermeneutics, II*] (1986), and some other relevant writings.[29]

Ricoeur's hermeneutics of the text generally revolves around the conceptual pair of "belonging (*appartenance*)" and "distanciation." On the one hand, like Gadamer, Ricoeur basically prioritizes "belonging to a world" (1995, 34) over distance or distanciation. As he puts it, "all objectifying knowledge about our position in society, in a social class, in a cultural tradition and in history is preceded by a relation of *belonging* upon which we can never entirely reflect." While not uniquely emphasizing belonging to *tradition*, Ricoeur maintains on more or less Gadamerian lines that "[b]efore any critical distance, we belong to a history, to a class, to a nation, to a culture, to one or several traditions" (1986, 328/267).[30] On the other hand, Ricoeur does not accept a sheer "antinomy" or "alternative" between belonging and distanciation, which seems to be "the mainspring of Gadamer's work" (1986, 101/75). In Ricoeur's view, distanciation is not "simply alienating," as Gadamer's term *Verfremdung* suggests (1986, 51/35).[31] Distanciation rather has a "positive and productive function" (1986, 102/76)[32]: It not only cancels but also preserves a subjective, singular content in an objective, universal form (see Lawlor 1992, 54). The concept of distanciation thus constitutes "the dialectical counterpart of the notion of belonging" (1986, 51/35; cf. 59/41, 365/297).[33]

even after his 'hermeneutical turn,' Ricoeur stresses "a profound affinity between phenomenology and hermeneutics" and claims "adherence to a sort of hermeneutical phenomenology," while rejecting "the idealist interpretation that Husserl gave of [his phenomenology]" (1995, 34, 36).

29 *Du texte à l'action* is a collection of papers published earlier, from 1970 until the mid-1980s. Some of the English versions of these articles can also be found in *Hermeneutics and the Human Sciences* (1981).

30 For parallel passages, see Ricoeur (1986, 28/15, 33f./19, 181/143; 1981, 294). At one point, Ricoeur even characterizes distanciation itself as "a moment of belonging" (1986, 330/268). As regards the relation between understanding and explanation, he holds that "explanation is not primary but secondary in relation to understanding" (1986, 22/10; cf. 123). See Lawlor (1992, 53f.).

31 According to Ricoeur, this dichotomy between belonging and "alienating distanciation" prevents Gadamer from "recognizing the critical instance" (1986, 365/297).

32 As Ricoeur acknowledges, however, Gadamer's view of distanciation is not simply negative: Though insufficiently, Gadamer "put his finger on the central problem of distanciation, which is not only temporal distance [. . .], but positive distancing; *a consciousness exposed to the efficacy of history* can understand only under the condition of distance." Ricoeur considers his own approach as an effort to "push further in the same direction" (1986, 329/268). See Gadamer (1972, 281/297).

33 Indeed, in Ricoeur's view, hermeneutic philosophy itself carries out distanciation precisely in reflecting on our belonging to the world. As he puts it, "[h]ermeneutics begins when, not content to belong to the historical world considered in the mode of the transmission of tradition, we interrupt the relation of belonging in order to signify it" (1995, 36).

Within this overall dialectic of belonging and distanciation,[34] Ricoeur thematizes the "process of discourse" in terms of "the dialectic of event and meaning" (1986, 103/77; cf. 108/81). To begin with, he argues, discourse, in contrast to the system of language (*langue*), "is given as an event" (1986, 103/77; cf. 184/145). That is to say, first, "discourse is realized temporally and in the present, whereas the system of language is virtual and outside of time." Second, while the system of language has no subject, "discourse refers back to its speaker by means of a complex set of indicators, such as personal pronouns." The eventful character of discourse is thus "linked to the person who speaks." Third, while "the signs of language refer only to other signs in the interior of the same system," discourse "is always about something." That is, "[d]iscourse refers to a world which it claims to describe, express or represent."[35] Fourth, and finally, "while language is only a prior condition of communication [...], it is in discourse that all messages are exchanged." Discourse thus "has an other, another person, an interlocutor to whom it is addressed." In Ricoeur's view, "[a]ll these features, taken together, constitute discourse as an event" (1986, 104/78; cf. 185/146).[36]

Next, however, Ricoeur continues, "all discourse is understood as meaning" (1986, 105/78; cf. 1975, 92/70). As he puts it, "[w]hat we wish to understand is not the fleeting event, but rather the meaning which endures." In other words:

> [...] discourse, by entering the process of understanding, surpasses itself as event and becomes meaning. The surpassing of the event by the meaning is characteristic of discourse as such. It attests to the very intentionality of language, to the relation within the latter of the noema and the noesis. If language is a *meinen*, a meaningful intention, it is precisely in virtue of the surpassing of the event by the meaning (Ricoeur 1986, 105/78).[37]

Here there occurs the "first distanciation," namely "the distanciation of the saying in the said" (1986, 105/78; cf. 185/146, 369/299). This "detachment of meaning from the event" (1976, 25), continues Ricoeur, is fully manifested in "writing" or the written text, which "is, par excellence, the basis for communication in and through distance" (1986, 51/35). First, by fixing discourse in an "exterior bearer," writing fixes "the meaning of the speech event, not the event as event" (1976, 26f.; cf. 1986, 185/146). Second, as a result, "the verbal meaning of the text" is dissociated from "the mental intention of the author," so that the text gains "semantic autonomy"

---

[34] Ricoeur also speaks of "the dialectic of participation and distanciation" (1986, 99/73; cf. 101/76).

[35] This point is related to the distinction Ricoeur draws (following Frege) between the "sense" and the "reference" of a proposition: While "the sense is the ideal object that the proposition intends and hence is purely immanent in discourse," "[t]he reference is the truth value of the proposition, its claim to reach reality" (1986, 113/85).

[36] This is broadly in accordance with Ricoeur's general fourfold formula of discourse: "someone says something to someone about something" (1986, 110/83; cf. 1995, 22, 24).

[37] What Ricoeur means here by the (Husserlian) relation between noesis and noema is more or less equivalent to the relation between the utterer's meaning and the utterance meaning or between "the instance of discourse" and "the 'intended' of discourse," which he mentions elsewhere (1975, 93/70). See also Lawlor (1992, 56).

(1976, 29). Third, writing also abolishes "the ostensive character of reference," and instead opens up the possibility of reference on another level, namely, reference to "the world of the text" (1986, 114f./85f.; cf. 188f./149). Fourth, the written text is addressed no longer to a particular person, but "potentially to whoever know how to read," that is, to a "universal audience" (1976, 31; cf. 1971, 183; 1986, 189/150, 198/157).[38] In this way, writing fully manifests "something that is in a virtual state, something nascent and inchoate, in living speech, namely the detachment of meaning from the event" (1976, 25).[39]

We have so far seen two stages of the dialectic process of discourse – the first in which discourse is given as an event, and the second in which it is distanced as meaning, especially through the mediation of writing. This process does not, however, end here. Rather, continues Ricoeur, there arises the third and final stage in which "the meaning surpasses itself in a new event of discourse, which is interpretation itself" (1971, 183). No less important is his point that interpretation, the final phase of the dialectic of event and meaning, proceeds itself dialectically.[40] In Ricoeur's account, "the dialectic of event and meaning, so essential to the structure of discourse, [. . .] generates a correlative dialectic in reading between understanding or comprehension [. . .] and explanation," and this dialectic constitutes interpretation.[41] Although understanding and explanation tend to become distinct poles, this polarity "must not be treated in dualistic terms, but as a complex and highly mediated dialectic" (1976, 74).

Ricoeur describes this dialectic of explanation and understanding as a sequence of two sub-processes: "a move from understanding to explaining" and "a move from explanation to comprehension." In the first move, he argues, interpretation starts with a "guess" or "a naive grasping of the meaning of the text as a whole" (1976, 74). Understanding must initially take the form of a guess owing to "the disjunction of the verbal meaning of the text from the mental intention of the author" (1976, 75). For the same reason, however, the initial guess needs to be subsequently tested and 'validated' by a procedure of explanation (1976, 78). These two factors, guess and validation, are "in a sense circularly related as subjective and objective approaches

---

[38] See Thompson (1981, 52).

[39] In this sense, according to Ricoeur, distanciation is not "something superfluous and parasitical," but rather "is constitutive of the phenomenon of the text as writing" (1986, 112/84).

[40] Etsurō Makita points to what he sees as an ambiguity in Ricoeur's account of the relation between the process of discourse and that of interpretation: On the one hand (as I have just noted in the text), Ricoeur designates interpretation as the third and last stage of the process of discourse. On the other, in different contexts, he supposes the circumstance that interpretation corresponds to the whole process of discourse, and, more specifically, that the three stages of interpretation (naïve grasp, explanation, critical understanding) correspond, respectively, to the three stages of discourse (event, meaning, new event) (see Makita 1997, 252ff.). See also Ricoeur (1976, 71f.).

[41] In his texts from 1971 onward, Ricoeur thus applies the term 'interpretation' "to the whole process that encompasses explanation and understanding" or, in other words, to the "dialectic of explanation and understanding" (1976, 74). Earlier, however, by interpretation he had meant a particular case of understanding, namely, the understanding of the text. See Makita (1997, 169f.).

to the text," and this relation constitutes a necessary part of the dialectical process of interpretation (1976, 79).[42]

In Ricoeur's view, this "dialectic between understanding as guessing and explanation as validation" is followed by the second move of the same dialectic, which proceeds, however, "in the reverse order" (1976, 80).[43] In order to show how "explanation [...] requires understanding," he focuses on a specific modern form of explanation, "structural analysis."[44] For example, "even in the most formalized presentation of myths by Lévi-Strauss, the units, which he calls mythemes, are still expressed as sentences, which bear meaning and reference." Structural analysis thus "does not exclude, but presupposes" the meaning to be understood (1976, 86). Furthermore, structural analysis "lead[s] us from a surface semantic [...] to a depth semantic," that is, to the level of the "ultimate 'referent'" of the text (1976, 87). As Ricoeur puts it:

> If [...] we consider structural analysis as one stage – albeit a necessary one – between a naïve interpretation and a critical one, between a surface semantic and a depth interpretation, then it would be possible to locate explanation and understanding at two different stages of a unique hermeneutic arc (Ricoeur 1976, 87).

In other words, the dialectic of understanding and explanation finally returns to understanding, yet this understanding is "critical" because it has already been mediated by explanation.[45] In this way, the dialectical process of interpretation is characterized not by "the opposition [...] between explanation and understanding," but by the "complementarity" between the two (1986, 75/53; cf. 1981, 183).[46]

---

[42] Ricoeur not only maintains that "the hermeneutical circle remains an unavoidable structure of interpretation" (1981, 178; cf. 171; 1976, 77; 1986, 200/158), but broadly follows Heidegger and Gadamer in displacing the hermeneutical circle "from a subjectivistic level to an ontological plane" (1981, 178). Closely linking the term 'circle' to the term 'dialectic,' Ricoeur generally considers "the correlation between explanation and understanding" to be "ultimately" the "hermeneutical circle" (1981, 221; cf. 1986, 211/167). More specifically, however, the circle appears at several junctures of the process of interpretation. For example, there is a circle "between my mode of being – beyond the knowledge which I may have of it – and the mode opened up and disclosed by the text as the world of the work" (1981, 178).

[43] According to Ricoeur, while the first move of the dialectic is "roughly the counterpart of the dialectic between event and meaning," the second move "may be related to another polarity of the structure of discourse, that of sense and reference" (1976, 80).

[44] For Ricoeur's critical engagement with structuralism during the 1960s, see Ricoeur (1969, 31ff./27ff.). See also 1995, 19.

[45] Ricoeur expresses the important role of explanation by means of the formula: "explaining more in order to understand better" (1995, 31). With regard to psychoanalysis, for instance, he holds that "an explanation by means of causes" is called for "in order to reach an understanding in terms of motives" (1981, 263).

[46] In his 1970 article "Qu'est-ce qu'un texte?" (included in 1986, 137–59/105–24), however, still retaining the narrow concept of interpretation, Ricoeur speaks of the complementarity between explanation and *interpretation* (rather than understanding) (1986, 142/110, 151/118, 154/120).

In Ricoeur's account, the final stage of the above dialectic, critical understanding, appears as "appropriation" (1981, 185). While traditionally (and in itself correctly) characterized as the "application (*Anwendung*) of the text to the present situation of the reader," appropriation may be reconceived as a "response" to the distanciation characteristic of explanation (1986, 116/87, 54/37). Specifically, it would be misguided to consider appropriation to be "a kind of possession" of the meaning by the present reader (1981, 192; cf. 1976, 93f.). Rather, "what is appropriation from one point of view is disappropriation from another."[47] To be sure, "[t]o appropriate is to make what was alien become one's own," and "what is appropriated is [. . .] the matter of the text" or "the world of the work" (1986, 54/37, 117/87). However, "the matter of the text becomes my own only if I disappropriate myself, in order to let the matter of the text be" (1986, 54/37). That is, "[i]t is not a question of imposing upon the text our finite capacity for understanding, but of exposing ourselves to the text and receiving from it an enlarged self." In this sense, "[a]s reader, I find myself only by losing myself" (1986, 117/88) – a circumstance that may be "expressed as a distanciation of self from itself within the interior of appropriation" (1986, 54/37).[48] In Ricoeur's view, appropriation is thus "dialectically linked to distanciation" (1986, 116/87; cf. 370/301; 1976, 89) and stands in a "complementary" relation to the objectification of meaning (1981, 185; cf. 183).[49]

So far I have outlined Ricoeur's hermeneutics of the text as it revolves around the dialectic of belonging and distanciation. After developing this theory largely during the early 1970s, Ricoeur extended it to the problem of human action, a central thematic of the social sciences. Here, however, I must be content to note simply that he approached this thematic of action in close analogy with the hermeneutics of the text, namely, in terms of the dialectic relation between event and meaning as well as between understanding and explanation (see 1986, 183ff./144ff.; 1995, 31).[50] In other words, his hermeneutic account of the text and its interpretation served as a paradigmatic theory for his subsequent project of a hermeneutics of action.

---

[47] Ricoeur warns also against some other misunderstandings of appropriation. In his account, appropriation does not imply "a direct congeniality of one soul with another" (1981, 191; cf. 1976, 92), but rather is "understanding at and through distance" (1986, 116/87). Nor is appropriation "governed by the original audience's understanding," but leaves the meaning of a text "open to anyone who can read" (1981, 192; cf. 1976, 93).

[48] In Ricoeur's account, while the other is not altogether inaccessible, "the tension between the other and oneself is unsurpassable." Therefore "[w]e exist neither in closed horizons, nor within a horizon that is unique." Ricoeur is critical of Gadamer's conception of the horizon insofar as the latter "seems to accept, at one stage, the idea of a single horizon encompassing all points of view" (Ricoeur 1986, 348/282). See Gadamer (1972, 288/304).

[49] It is worth touching on Ricoeur's view of the limited "univocity" of interpretation. According to Ricoeur, interpretation "consists in recognising which relatively univocal message the speaker has constructed on the polysemic basis of the common lexicon" (1986, 77/55). While interpretation is "a struggle for univocity," one cannot attain absolute univocity (1975, 383). For this reason, "the conflict of interpretations is insurmountable and inescapable," and "absolute knowledge is impossible" (1981, 193). See also Lawlor (1992, 59ff.).

[50] See Thompson (1981, 63f.).

The previous and the present sections have been devoted to a review of recent and contemporary hermeneutic philosophy as it is represented by the work of Gadamer and Ricoeur. This survey serves as a starting point for my subsequent inquiry into the possible conceptual connections with Bohr's thought. To be sure, as we can readily see, Ricoeur's hermeneutics is no more directly or thematically concerned with natural science than Gadamer's. The focus of these philosophers' work continues to be on linguistically-mediated interpersonal understanding, whereas Bohr's complementarity pertains to non-verbal and material intervention in nature as coupled with the use of language. Despite this difference, based on their difference in object domain, however, there will prove to be important similarities between their philosophical orientations. In the next section, with a focus on Gadamer's conceptual pair of belonging and alienation as well as Ricoeur's dialectic of belonging and distanciation, I will examine how these hermeneutic ideas are parallel to – while at the same time different from – Bohr's complementarity as the 'actor–spectator' relationship.

5.3 Complementarity and Hermeneutic Philosophy

There is little documentary evidence that Bohr was ever influenced by the hermeneutic philosophy of his times or earlier periods. It appears, however, that, through Høffding's mediation, he familiarized himself with the notion of the hermeneutic circle, and that he perhaps considered this notion to have something to do with his own idea of complementarity. In his 1931 "Tribute to the Memory of Harald Høffding," he describes the late philosopher's account of "the phenomena of consciousness" as follows.

> A prominent feature is the striving to maintain the balance between analysis and synthesis, and it is never forgotten that although the whole is composed of individual parts, these in their turn appear in the light of the whole (NBCW, 10:321).

This not only attests to Bohr's knowledge of the idea, if not the term, of the hermeneutic circle of part and whole, but also suggests its connection with the conceptual pair of analysis and synthesis, which he himself employed in his account of complementarity.[51] We can thus see a possible – though not further specified – conceptual linkage between the hermeneutic circle and Bohr's complementarity. In this light, von Weizsäcker's interpretation of Bohr's thought in terms of "circular complementarity" as reviewed in Section 3.1 is by no means an arbitrary reading from without, but seems to do justice to the above possible relevance of the hermeneutic notion of the circle.

This idea of the circle – mentioned and perhaps also embraced by Bohr as well as by von Weizsäcker – may be characterized, however, as a 'classical' concept of the hermeneutic circle, or what Gadamer calls "a formal relation between part and whole." That is to say, in Heideggerian terms, it is not concerned with

---

[51] In Section 2.3 (note 59), I commented on Bohr's notions of analysis and synthesis in his complementarity argument.

the 'ontological' conditions of understanding or interpretation, but rather with the 'ontic' structure of the object interpreted, or with the corresponding procedure through which interpretation is carried out. Yet, as is suggested by our investigation up until now, especially in Chapter 4, Bohr's complementarity in effect goes beyond this pre-Heideggerian hermeneutic framework. While his stated notion of the whole-part relation as seen above is restricted to the classical concept of the hermeneutic circle, his very idea of complementarity, which revolves around the relation between the roles of 'actors' and 'spectators,' indicates its possible connections with more recent ideas of hermeneutic philosophy. In what follows, I will accordingly examine the conceptual intersections of Bohr's complementarity with post-Heideggerian hermeneutic philosophy, specifically with the thought of Gadamer and then of Ricoeur.

As we saw in Chapter 4, it is Bohr's basic idea that as 'actors' we participate or are involved in the phenomena, whether physical, biological, or psychical. Notwithstanding the term 'actor,' it is not the case that the subject one-sidedly influences the object, but that the two sides *interact* with each other.[52] Furthermore, this interaction is such that the two sides do not exist separately, but are essentially and from the outset interdependent. That is, the 'actor' does not simply interact with the object from without, but is involved in the situation in which there is no "independent reality" of either subject or object. Let us compare this notion of the 'actor's' interacting involvement with the hermeneutic concept of belonging.

Hans-Georg Gadamer's 'philosophical hermeneutics' as reviewed in Section 5.1 is centrally concerned with the idea of "belonging," specifically belonging to a tradition. In Gadamer's view, belonging to a tradition constitutes a prior condition of the understanding of the past, because understanding "proceeds from the commonality that binds us to the tradition." Even though the term 'belonging' bears the connotation of passive submission or bondage, he also holds that "[t]radition is not simply a permanent precondition," but that we "participate in the evolution of tradition, and hence further determine it ourselves" (1972, 277/293). To this extent, namely, insofar as belonging is not a simple subjection to the "historically given," but is characterized by "the interplay of the movement of tradition and the movement of the interpreter" (1972, 265/281, 277/293), we can speak of its similarity to Bohr's notion of the 'actor's' interacting involvement. It is also worth recalling Gadamer's account of belonging in the aesthetic sphere, as illustrated by the example of the theatrical play, which bears a literal resemblance to Bohr's metaphor of 'actor' and 'spectator.'

In contrast to the 'actor,' Bohr's 'spectator' leaves the object intact, thus maintaining the subject/object separation. In so doing – typically in the case of the claim of causality in quantum theory – the spectator preserves the 'quasi-omnitemporal' or 'ideal' character of the object. This detachment of the 'spectator' from the object

[52] While concerned with the "subjective" conditions of our experience, Bohr's thought may be called no more subjectivist than Gadamer's or Ricoeur's, despite misunderstandings in this regard among some commentators. See Section 3.2.

appears to be more or less parallel to the hermeneutic conception of alienation or distanciation. As we have seen, Gadamer holds that, despite the importance of belonging, our relation to a tradition is not an unquestionable unity with it, but contains "a polarity of familiarity and strangeness" or of proximity and distance (1972, 279/295). This problem of distance becomes most manifest with regard to "writing" as "a kind of alienated speech" (1972, 371/393). We can thus see how the detached character of Bohr's 'spectator' corresponds to Gadamer's concept of alienation, associated with the notion of writing as "the abstract ideality of meaning" (1972, 370/392). Taken together with the above point on the 'actor,' this suggests the following analogy between the two thinkers' overall approaches. Bohr's complementarity may generally be characterized as a project of questioning the traditional notion of the 'spectator's' purely detached knowledge and of indicating how this notion is undermined by the irreducible moment of the 'actor's' involvement. To this extent, it is parallel to the basic orientation of Gadamer's hermeneutics, which seeks to overcome alienation by virtue of the irreducible moment of belonging (see Gadamer 1976, 4ff.).[53]

As we saw in the previous section, Paul Ricoeur elaborated the relation between "belonging" and "distanciation" from a hermeneutic point of view related to, yet distinct from, Gadamer's. As it will turn out, Ricoeur's hermeneutics of the text may in certain respects be connected even more closely with Bohr's complementarity. According to Ricoeur, distanciation is not "simply alienating," but rather has a "positive and productive function," and thus stands in a "dialectical" relation to belonging (1986, 51/35, 102/76). In his account of discourse, the dialectic of belonging and distanciation takes the form of "the dialectic of event and meaning." Discourse is given as an event, but then "surpasses itself as event and becomes meaning," which in turn "surpasses itself in a new event of discourse," that is, interpretation (1986, 103/77, 105/78). Further, this interpretation as the final stage of discourse proceeds itself dialectically, and this dialectic is characterized not by the opposition, but by the "complementarity," between understanding and explanation (1986, 75/53). This suggests a close conceptual linkage between Bohr's and Ricoeur's ideas: What Bohr calls the 'actor' is parallel to Ricoeur's concept of belonging, while the former's notion of the 'spectator' corresponds to the latter's distanciation. In this way, Bohr's complementarity of the roles of 'actors' and 'spectators' is analogous to Ricoeur's dialectical relation between belonging and distanciation, or, in more specific contexts, between event and meaning as well as between understanding and explanation. The conceptual correspondence between

[53] This is not to say, however, that all theoretical elements of Bohr's complementarity argument have exact counterparts in Gadamer's hermeneutics or vice versa. Specifically, although the phenomenon in Bohr's account of quantum theory seems to correspond broadly to Gadamer's notion of the text, Bohr focuses on our involvement in individual phenomena, whereas Gadamer emphasizes our belonging to the whole tradition through which the text and its subject matter are handed down to us. Later in this section, I will discuss further conceptual differences between Bohr's complementarity and hermeneutic philosophy.

Table 5.1 Bohr's complementarity and hermeneutic philosophy

Bohr	'Actor'	'Spectator'
Gadamer	Belonging	Alienation
Ricoeur	Belonging	Distanciation
	Event	Meaning
	Understanding	Explanation

Bohr's complementarity and the hermeneutic philosophy of Gadamer and Ricoeur as indicated so far may be succinctly expressed by Table 5.1.[54]

For a further elaboration of this relationship, however, we must attend to different layers and phases of thought in Bohr's complementarity as analyzed earlier. In Chapter 4, I distinguished the following three conceptions of complementarity: the two introduced in the 'early' period – static-contrastive juxtaposition and dynamic displacement – and the one developed from the 'middle' period onward, static-symmetrical juxtaposition. In what follows, based on my earlier analysis of these conceptions, I will examine whether and, if so, how each of them may be connected with Ricoeur's hermeneutic ideas. Where appropriate, I wish to make supplementary comments on the relation to Gadamer's thought as well.

(1) As we have seen, Bohr's 'early' conception of complementarity as *static-contrastive juxtaposition* is characterized by the circumstance that one behaves as an 'actor' and as a 'spectator,' not at the same time or in the same situation, but only in separate and incompatible situations. This is exemplified by the relation in quantum theory between space-time coordination and the claim of causality, or between observation and the definition of the state. Here observation may be considered as a kind of 'event' in Ricoeur's sense, and the definition of the state as a determination of 'meaning' with regard specifically to the momentum or energy of the atomic object. This enables us to see for the moment how Bohr's complementarity in its static-contrastive version may be likened to Ricoeur's relation between event and meaning.

In this conception of complementarity, however, the roles of the 'spectator' and the 'actor' simply stand side by side, without undergoing any movement or process of transition between them. In contrast, Ricoeur's hermeneutics concerns precisely the dynamic process of discourse and interpretation, that is, the process of transition from belonging to distanciation and vice versa – specifically, from discourse as an event to discourse as meaning and thence further to a new event of discourse.

[54] A brief comment may be in order on this table, with regard specifically to the section for Ricoeur. The two-column arrangement of Ricoeur's paired terms expresses his idea that the relation between belonging and distanciation appears – in the process of discourse – as the relation between event and meaning, while, in interpretation, taking the form of the relation between understanding and explanation. It does *not* mean, however, that event and meaning correspond directly to understanding and explanation, respectively, in such a way, for example, that one understands the event while one explains the meaning. Admittedly, as noted in the previous section (note 40), Makita points out that Ricoeur's account leaves a certain ambiguity as to the relationship among the above key terms.

This is broadly a feature common to Gadamer's philosophy as well: He discusses the process by which the meaning of spoken language is alienated in writing, and then the reverse process through which this alienation is overcome in interpretation. In this way, Bohr's static-contrastive conception of complementarity is devoid of the dimension of process or movement crucial to Gadamer's as well as Ricoeur's hermeneutic philosophy, so that their conceptual similarity is considerably limited.

(2) Before proceeding to Bohr's dynamic conception of complementarity, I will briefly comment on his other static notion, the *static-symmetrical* conception, developed from the 'middle' period onward. Not only does this conceptual scheme also lack the dynamic dimension, but, unlike the above static-contrastive conception, it juxtaposes two complementary relata – such as the two types of phenomena corresponding to space-time coordination and momentum-energy conservation – as similar and interchangeable items. Here complementarity is no longer conceived as a relation of contrast between the 'actor's' involvement and 'spectator's' detachment. We can readily see that this conception is quite foreign to the hermeneutic relation between belonging and alienation or distanciation. Furthermore, it is based on this static-symmetrical conception that Bohr in the 'late' period turned toward an objectivist philosophy. Marked by the term "detached observer," implying a pure 'spectator' free of the moment of being an 'actor,' this objectivist tendency is obviously incompatible with hermeneutic philosophy in its either Gadamerian or Ricoeurian form. From this point of view, Bohr's philosophical path from the 'middle' to the 'late' period may be characterized as a shift from a quasi-hermeneutic view of knowledge toward a form of scientific rationalism.

(3) Bohr's conception of complementarity as *dynamic displacement* was initially largely restricted to fields outside physics, but in the 'middle' period carried over into quantum theory, in which it came to play a crucial role. If we first take up his idea of complementarity in epistemology, which revolves around the relation between the application of a concept and its conscious analysis, we can perceive a direct and obvious connection with Ricoeur's hermeneutic account. That is, the application of a concept may be viewed as an event of discourse, while its conscious analysis may be characterized as an analysis of meaning or, more precisely, as an analysis of the event *as* meaning. Further, in Bohr's view, this analysis of meaning turns out to involve another application, an application of another concept or set of concepts. This may be compared with Ricoeur's third stage of the discursive process, in which "the meaning surpasses itself in a new event of discourse." This close affinity with Ricoeur's hermeneutic theory also applies to Bohr's reinterpretation of quantum theory in the 'middle' period. In this interpretation, any measurement serves as the determination of a 'meaning' regarding either space-time coordination or momentum-energy conservation, and yet proves unavoidably to have the character of an 'event' involving an interaction between object and instrument. In this way, Bohr's dynamic conception of complementarity in quantum theory as well as other fields is found to be in close proximity to Ricoeur's dialectic of event and meaning.[55]

[55] Bohr's dynamic conception of complementarity may also be compared with Ricoeur's account of interpretation in terms of the "complementarity" of understanding and explanation. Bohr's

Even at this closest point of convergence, however, there is a subtle yet crucial difference between the two thinkers' approaches. As we have seen, in Bohr's dynamic conception, the analysis of the application of a concept turns out *itself* to be another application, and – in the context of quantum theory – a measurement as the determination of a 'meaning' appears *itself* as a phenomenal 'event.' By contrast, in Ricoeur's approach, discourse as meaning and interpretation as a new event are situated as two consecutive phases or stages. More generally, Bohr's dynamic conception holds that the 'spectator's' reflection on the 'actor's' involvement proves to have itself a character of involvement, whereas, in Ricoeur's account, distanciation *goes over into* a stage of belonging. As a result, Bohr's notion of the indefinitely extensible series of reflections, which implies the undecidability between the alternating roles of 'actors' and 'spectators,' considerably differs from Ricoeur's concept of dialectics, whose two moments are each determined as such and are put in order as definite stages to form what is called the "hermeneutical arc" (1976, 87).

We have seen not only striking similarities, but also a hardly ignorable difference, between Bohr's dynamic conception of complementarity and Ricoeur's hermeneutics. It seems to me that parallel points may be made on the relation between Bohr's and Gadamer's thought as well, although this case does not lend itself to an analysis as systematic as in the above case with Ricoeur's ideas. We can see that, among Bohr's different conceptions of complementarity, the dynamic conception may be the most closely connected with the hermeneutical relation between belonging and distanciation, but that it nevertheless does not fully converge with the latter.

There are further important differences between Bohr's complementarity and hermeneutic philosophy. First, Bohr's emphasis on the irreducible moment of the 'actor' does not mean that he prioritizes the 'actor's' involvement over the 'spectator's' detachment. In Bohr's static-contrastive conception, two complementary relata such as space-time coordination and the claim of causality stand obviously on an equal footing. Also in his dynamic conception, the 'actor's' involvement is not ontologically privileged over the 'spectator's' reflection, although the latter is reflection *upon* the former. More specifically, as we have seen earlier, complementarity in fields outside physics indeed has a series of reflections starting from the 'actor's' involvement, and yet this starting point does not constitute an absolute beginning in the ontological sense. Furthermore, Bohr's 'middle'-period dynamic idea of complementarity in quantum theory develops a similar but somewhat different series of reflections which, as explicitly shown by Table 4.5, has *no* definite beginning. In contrast, hermeneutic philosophy as reviewed earlier generally gives priority to belonging over alienation or distanciation. This is most manifest in the

---

notion of the 'spectator's' detached reflection, which objectifies the meaning experienced by the 'actor,' more or less corresponds to what Ricoeur characterizes as the transition from naive understanding to explanation. Further, Bohr's subsequent recomprehension of reflection as yet another involvement may be regarded provisionally as similar to Ricoeur's notion that explanation leads to critical understanding or appropriation. There seems to arise a difference, however, between these ideas of the two thinkers – a difference parallel to the one pointed out subsequently in the text with regard to Ricoeur's dialectic of event and meaning.

case of Gadamer's theory, according to which our original belonging to a tradition is characterized by "the living present of conversation," and the interpretation of texts lies in a transformation of alienated written signs back into "the actual language of speech" (1972, 350/368, 371/393). While rejecting the idea of "the reproduction of the original production" as conceived by Romantic hermeneutics (1972, 159/167), Gadamer nevertheless regards the interpretive process as directed toward an overcoming of alienation and a full recovery of belonging. Although Ricoeur's hermeneutics differs from Gadamer's in emphasizing the "positive and productive" character of distanciation, he also holds that "all objectifying knowledge [. . .] is preceded by a relation of belonging" (1986, 102/76, 328/267). This enables us to see how Bohr's notion of the 'actor–spectator' relationship diverges from Gadamer's as well as Ricoeur's conception of belonging as primary and original as against alienation or distanciation.

Second, in view of Ricoeur's use of the term 'complementarity' in his account of understanding and explanation, it is worth comparing the meanings he and Bohr assign to this term. As we saw in Chapter 2, an important characteristic of Bohr's concept of complementarity is that it means not only 'joint completion,' but also 'mutual exclusion.' In Section 4.1, I discussed how this latter notion, mutual exclusion, is constitutive not only of his static conception of complementarity, but – though in a different way – of his dynamic conception as well. In contrast, when Ricoeur uses the term 'complementarity,' it is largely in its ordinary sense of joint completion *as against* the "exclusive alternative" between different elements (1986, 142/110). He maintains, in particular, that "the opposition, disastrous in my view, between explanation and understanding" should be replaced by "a complementarity between these two attitudes" (1986, 75/53). His notion of the complementarity of understanding and explanation is thus oriented toward a "reconciliation" of the two through dialectical mediation (1986, 155/121).[56] We can thus see how Bohr's concept of complementarity differs from Ricoeur's, the latter devoid of, and directed against, the notion of mutual exclusion.

In this section, devoted to an analysis of the conceptual intersections between Bohr's complementarity and hermeneutic philosophy, I have indicated how the two sets of philosophical approaches exhibit close affinities, while at the same time they remain heterogeneous in some other respects. Here we may also note that, owing to their difference in object sphere as noted earlier, the relationship between complementarity and hermeneutics just analyzed may be divided into the following two kinds of connections. On the one hand, insofar as Bohr concerns himself with the use and meaning of concepts and words, the relation of his thought to hermeneutics is not only analogous, but more substantive and internal. That is to say, the two philosophical orientations intersect with each other in their overlapping fields of inquiry. On the other hand, unlike the hermeneutic philosophers, the physicist Bohr

[56] Even as he remarks that "explanation and interpretation are indefinitely opposed and reconciled" (1986, 159/124), he does not seem to apply the term 'complementary' – in a manner similar to Bohr's usage – to the double sense of "opposed and reconciled."

focuses also – and indeed primarily – on our material and technical intervention in natural processes, and, to this extent, their conceptual links, however close they may be, are restricted to analogical ones. This restriction is, however, largely due to my choice of Gadamer's and Ricoeur's work as representative of contemporary hermeneutics – a choice which may not ultimately be justified.

So far in this chapter, I have not discussed other hermeneutic approaches such as Apel's transcendental hermeneutics or Heelan's phenomenological-hermeneutic philosophy of science, which in one way or another address the question of natural science and its material-technical engagement with nature. As a partial corrective to the above restriction, I wish accordingly to conclude the chapter by briefly commenting on Apel's and Heelan's work and their possible connections with Bohr's complementarity.

(1) Karl-Otto Apel's philosophical program designated as "transcendental hermeneutics" (1973, 2:187/100) diverges in certain respects from mainstream hermeneutic philosophy.[57] In his major 1973 work *Transformation der Philosophie* [*Towards a Transformation of Philosophy*], Apel develops a new "theory of science" by enlarging "the Kantian question of the 'preconditions for the possibility of knowledge'" (1973, 2:96/46). In his account, human knowledge rests not only on "the a priori of consciousness (*Bewußtseinsapriori*)," as it has been thematized in traditional epistemology, but also on another a priori that he names "the bodily a priori (*Leibapriori*)." While the a priori of consciousness constitutes the precondition of "knowledge through reflection," the bodily a priori is the precondition of "knowledge through engagement." Apel maintains that "the bodily a priori of knowledge stands in a complementary relationship to the a priori of consciousness," in the sense that the two preconditions "supplement each other" while in actual situations either one of them "takes up the leading position" (1973, 2:99/48f., trans. mod.).

Further, continues Apel, the second type of knowledge, namely knowledge through bodily engagement, is divided into two subtypes, and each of them is related to a specific a priori "cognitive interest." In his view, "man has basically two equally important but not identical *complementary* cognitive interests" (1973, 2:112/59). One of them is the "technical" interest or the "interest in technically relevant knowledge of nature," which underlies the natural sciences and is particularly constitutive of the "experimental engagement of modern physics" (1973, 2:122/67, 100/49). The other is the "hermeneutic" interest, namely, the "interest in intersubjective agreement about possible interpretable motivations in life," which lies at the basis of the human sciences or humanities (1973, 2:122/67).

According to Apel, thus based on the two complementary cognitive interests, the "explanatory natural sciences" and the "interpretative human sciences" are themselves complementary to each other (1973, 2:101/50; cf. 114/60). On the one

[57] I cannot enter into the work of Jürgen Habermas, who started from a philosophical point of view close to Apel's, and whose subsequent debate with Gadamer, in particular, has important bearing on contemporary hermeneutic philosophy and beyond. For a critical engagement with Habermas's thought in the context of natural science, see Radder (1988).

hand, he concurs with Gadamer in stressing the importance and indispensability of hermeneutic understanding, and criticizes, in particular, the neopositivist idea of a "unified science" with its attempt to reduce understanding to causal explanation. On the other hand, he points to certain limits of Gadamerian hermeneutics, arguing that our understanding encounters "contradictions in the expressions of life" that are not accessible to hermeneutic understanding, but "can only be analyzed by means of a quasi-objective explanatory science" (1973, 2:122f./68). In Apel's account, these two types of inquiry – "hermeneutic inquiry" and "natural scientific objectification and explanation" – stand in a "complementary relationship" in the sense that they are "mutually exclusive and yet none the less thereby supplement each other" (1973, 2:111/58, trans. mod.).[58]

We can see that, in certain respects, Apel's transcendental hermeneutics as just sketched comes close to Bohr's complementarity, perhaps partly even closer than Gadamer's and Ricoeur's theories do. First, Apel's idea of the complementary relation between the a priori of consciousness and the bodily a priori as well as between explanation and understanding converges to a large extent with Bohr's complementarity as a 'spectator–actor' relationship. It is particularly noteworthy that he explicitly includes *natural* science in speaking of "knowledge through engagement" based on the bodily a priori, which is parallel to the actor's involvement in Bohr's thought. Second, unlike Ricoeur, Apel uses the very term 'complementary' not in its ordinary sense, but precisely in the Bohrian double sense of mutual exclusion and joint completion. Third, again similarly to Bohr and unlike Gadamer and Ricoeur, he does not prioritize engagement over reflection, or understanding over explanation, but places them on an equal footing.

These conceptual affinities are, however, considerably limited by the following points: Apel basically restricts the theme of hermeneutic understanding to the humanities and some aspects of the social sciences. To be sure, he does not deny the intersubjective dimension of natural science, but holds, in particular, that "the natural scientists' community of experiment always expresses a semiotic community of interpretation" (1973, 2:112/58). Nevertheless, in Apel's view – in virtual agreement with the neopositivist view of science – natural-scientific knowledge as such is thoroughly oriented to "objectification and explanation" as against hermeneutic understanding.[59] Correspondingly, in his account of cognitive interests, he regards the technical, and not the hermeneutic interest as underlying natural science. Despite his attention to the bodily a priori for the natural as well as the human sciences, he

---

[58] Apel designates his own theoretical approach, thus oriented to "a dialectical mediation of 'explanation' and 'understanding,'" as "the critique of ideology" (1973, 2:101/50).

[59] At the first conference of the International Society for Hermeneutics and Science, held in Veszprém in September 1993, there arose an intense debate on the relevance of hermeneutics to natural science. As Dagfinn Føllesdal summarizes, Apel argued that hermeneutics applies to the studies *of* science but not studies *in* science, while Don Ihde and others contended that it does have a legitimate place in natural science itself. See Føllesdal's "Introduction" to Fehér et al. (1999, vii–xi, on viif.). With reference to this debate, Ihde (1998, 40) critically comments on Apel's view. See also Martin Eger, "Language and the Double Hermeneutic in Natural Science," in Fehér et al. (1999, 265–80, on 276).

does not pursue this theme in connection with, say, the possible limits of detached objectification in natural science, around which Bohr's complementarity argument revolves. This indicates how Apel's transcendental hermeneutics, while in some respects parallel to Bohr's complementarity, is nevertheless otherwise dissimilar to the latter.

(2) Let us now turn to Patrick A. Heelan's phenomenological-hermeneutic approach to science, developed in his major 1983 work *Space-Perception and the Philosophy of Science* and other related writings. Unlike Apel, Heelan does stress the hermeneutic dimension of natural science, and, from this point of view, specifically addresses quantum mechanics and complementarity.

Heelan's basic philosophical orientation may be briefly outlined as follows (see 1983, 266–70). First, as just noted, challenging the conventional tendency to limit the scope of hermeneutics to the humanities, Heelan argues that hermeneutics is no less relevant to natural science. In his account, "[t]he role of hermeneutics in natural science is not restricted to the study of literary and graphic materials," but is extended to the "reading" of "'textual' materials [. . .] made available by the use of appropriate readable technologies" (1983, 224; cf. 220, 269f.). Scientific observation – 'reading' a thermometer, for example – is analogous to reading texts written in natural languages, and is thus essentially "hermeneutical" (1983, 193; cf. 273f.). Second, Heelan postulates "the ontological primacy of perception," which means that "reality is exactly what is or could be manifested through perceptual essences and profiles as horizons of Worlds" (1983, 268; cf. 174, 192). Third, rejecting both "scientific realism" and "instrumentalism," Heelan advocates what he designates as "horizonal realism."[60] On the one hand, contrary to instrumentalism, horizonal realism holds that "science has the intent of describing the elements and the structures of reality" (1983, 269). On the other, it differs from scientific realism in "see[ing] reality from the start as 'horizonal'" (1983, 177). According to Heelan, "[t]o each horizon, there belongs a particular language and a corresponding *context* for its correct use," where the context is the horizonal structure itself that "has both subjective and objective components" (1983, 178).

From the above general philosophical point of view, Heelan specifically discusses what he calls "context-dependence" as it is exemplified by the conceptual structure of quantum mechanics. In his view, quantum mechanics is "the first natural science that in its explicit form includes reference to the contextual character of scientific inquiry" (1983, 208). Here he attaches special importance to the concept of "complementarity," which, in his account, is "a notion introduced by Niels Bohr to describe the relation between conjugate variables, such as position and momentum" (1977, 18). Taking the basic sense of the term 'complementarity' to be "context-dependence," he seeks to show that the logical structure of quantum mechanics

[60] Heelan defines scientific realism as "the belief that science has the power of uncovering *the real uniquely*," and instrumentalism as "the contrary belief that science does not concern itself with the real, but only with extending human power over nature" (1983, 18; cf. 173).

revolves around the relation between "complementary descriptive languages," and "this seems to have been the sense that Bohr himself had in mind" (1983, 179, 273; 1977, 18; cf. 1983, 184).[61] In this way, quantum mechanics, with the pivotal concept of complementarity, serves as a crucial scientific theory that paradigmatically supports his overall hermeneutic philosophy of science.

Heelan's hermeneutic theory as just sketched is commonly and rightly considered a pioneering work in the hermeneutics of natural science, and his account of quantum mechanics and complementarity, in particular, represents one of the few thematic inquiries into the subject from an explicitly hermeneutic point of view. Further, with his emphasis on instrumentation as "readable technology," Heelan's work has served as a starting point for the recent shift in the hermeneutics of science from the text-centered to the material and technological thematic (see Ihde 1991, 77ff.; 1998, 41).[62] This suggests the possibility of exploring Bohr's thought from a new hermeneutic perspective that focuses on the question of technology and material engagement with nature.

As regards the interpretation of Bohr's complementarity as such, however, Heelan's account appears to be problematic in crucial respects. His understanding of complementarity basically restricts itself to the relation between conjugate variables such as position and momentum, that is, to what von Weizsäcker refers to as 'parallel complementarity.' Heelan does not speak of an idea of complementarity similar or corresponding to von Weizsäcker's 'circular complementarity' or to what I have called the dynamic conception. Furthermore, he does not share other hermeneutic philosophers' concern with the relation between belonging and distanciation, and correlatively fails to attend to Bohr's notion of the 'actor–spectator' relationship, which I consider to be a crucial link with post-Heideggerian hermeneutic philosophy. For these reasons, while offering a significant hermeneutic perspective on modern physical science, Heelan's work does not seem to contribute essentially to our understanding of the relation between complementarity and hermeneutics as discussed in this chapter.

---

[61] Heelan makes the qualifying remark, however, that Bohr "might have had reservations about the full implications of the position I am attributing to him because of conflicts with his epistemology" (1977, 18; cf. 1970). In his early work *Quantum Mechanics and Objectivity*, he characterizes Bohr's philosophical position as "realist in the empiricist sense" (1965, 46).

[62] For a recent attempt to "expand the role of hermeneutics into technoscience," see Ihde (1998, esp. 137). In this context, we can also refer to Hans Radder's (1988) philosophical approach to physical science, specifically to quantum mechanics, which, by critically appropriating Jürgen Habermas's views on natural science, discusses the theme of experimentation as "material realization." In his more recent works (1996, 2002), notably in *The World Observed/The World Conceived* (2006), Radder focuses on the local realizations of observational processes and the nonlocal meanings of the results of such processes, which, taken together, may be compared with Ricoeur's paired notions of belonging and distanciation. It is also noteworthy that this 2006 work by Radder contains a favorable but partly critical assessment of Heelan's hermeneutics of natural science (2006, 57–70).

If we take account of yet other approaches of hermeneutic philosophy, we might be able to make further comparisons with Bohr's complementarity. In the present work, however, I will not move further in this direction, but rather – in the next and final chapter – take up another contemporary philosophical orientation which might be called "radicalized hermeneutics," namely, Derridean deconstruction.[63]

[63] This characterization of deconstruction as "radicalized hermeneutics" is Tetsuya Takahashi's, based on his view that a radicalization of "the hermeneutics of the text" would lead to an abandonment of "all nostalgia for presence" (1992, 238).

Chapter 6
Intersections with Derridean Deconstruction

In the recent and contemporary currents of thought commonly called postmodernism and poststructuralism, quantum theory has often been cited as an example of a new form of knowledge that supposedly goes beyond the conceptual framework of modern philosophy and science.[1] Until fairly lately, however, this line of characterization and invocation of quantum theory was in many cases hardly sustained by solid conceptual analysis of the subject.[2] It is only since the closing years of the twentieth century that quantum theory and, linked with it, Bohr's complementarity have become systematically explored with reference to the thematic of postmodernism, poststructuralism, or deconstruction.[3] These studies, including Michel Bitbol's[4] and Karen Barad's[5] inquiries, from their respective viewpoints shed new light on the philosophical problems of quantum theory, which hitherto

[1] Jean-François Lyotard regards "quantum mechanics and microphysics" – along with Gödel's theorem, fractal geometry, and catastrophe theory – as exemplifying what he calls "postmodern science" (1979, 88–97/53–60, esp. 91/56). See also Froula (1985).

[2] This seems to have constituted part of the background of what is known as the "Science Wars" during the 1990s. Lyotard's postmodernist account of quantum mechanics was criticized by Sokal and Bricmont (1998, 125ff.), in particular, from a rather narrowly physical-scientific point of view. I will briefly comment on the Science Wars toward the end of this chapter.

[3] Here I do not enter into the connections or differences between postmodernism, poststructuralism, and deconstruction – terms often carelessly confounded in the popular discourse on the subjects. Incidentally, Jacques Derrida never characterized himself as a poststructuralist, let alone as a postmodernist.

[4] Bitbol interprets Schrödinger's philosophical views on quantum mechanics not as "conservative" but rather as "postmodernist" (Bitbol 1996, 24f.), and, in this context, also compares Schrödinger's and Bohr's approaches (see 1996, 211ff.). Slavoy Žižek's account of quantum theory from the point of view of Lacanian psychoanalysis is also worth mentioning here, although it does not thematically address Bohr's thought (Žižek 1996, 189–236).

[5] In *Meeting the Universe Halfway* (2007) and other related writings (1996, 1998), Barad not only interprets Bohr's "philosophy-physics," but seeks to extend and partially revise his views in critical dialogue with current science studies and other interdisciplinary approaches such as feminist and poststructuralist theory. Starting from Bohr's account of the inseparability of the objects and the agencies of observation, she develops her own philosophical framework named "agential realism," which is a "nonrepresentationalist" form of realism that reconceives the material and the discursive, the natural and the cultural as "agentially intra-acting components" of the world (2007, 26, 56,

M. Katsumori, *Niels Bohr's Complementarity*, Boston Studies in the Philosophy of Science 286, DOI 10.1007/978-94-007-1748-0_6,

have not been adequately treated within the conventional frames of the history and philosophy of science. In particular, as far as Bohr's complementarity in relation to Derridean deconstruction is concerned, arguably the most elaborate work to date has been developed by Arkady Plotnitsky.

In the previous chapter, for want of relevant prior studies on the relation between complementarity and hermeneutics, I needed from the outset to develop my own interpretive point of view. In this chapter, by contrast, I wish to take up Plotnitsky's analysis of the intersections between Bohr's and Derrida's ideas as a suitable point of reference from which to examine further the possible links between the two thinkers' orientations. In the first and second sections, I outline Derridean deconstruction and Plotnitsky's account of the Bohr–Derrida relation, respectively. In Section 6.3, I examine a series of problems that seem to face, or be left unsolved by, Plotnitsky's analysis. This critical appraisal then leads me to present a new interpretive approach of my own in Section 6.4, which both extends and delimits the intersections between complementarity and deconstruction. In the course of this inquiry, both the structural and the historical complexity of Bohr's thought as discussed earlier will again become crucial. I conclude the chapter by briefly commenting on the possible implications of my inquiry here to some issues debated in what is called the "Science Wars."

6.1 Derrida's Project of Deconstruction

In this section, I review the philosophical thought of Jacques Derrida (1930–2004), which he himself calls, and is commonly known as, "deconstruction."[6] In so doing, I largely restrict myself to his earlier work – represented by *La voix et le phénomène* [*Speech and Phenomena*] and *De la grammatologie* [*Of Grammatology*], both published in 1967, among other writings from the 1960s and early seventies – which may be viewed as constituting a basic part of his work throughout his life.

Derrida's project of deconstruction is a kind of critical engagement with the whole tradition of Western philosophy stretching "from Plato to Husserl," or rather even "from the pre-Socratics to Heidegger" and beyond (1972c, 33/22; 1967b, 11/3). In his account, this philosophical tradition generally revolves around a series of conceptual oppositions such as "soul/body, good/evil, inside/outside, [...] speech/writing" to which one may further add the pairs of self/other, identity/difference, presence/absence, present/non-present, meaning/sign, truth/falsity, life/death, nature/culture, and so forth (1972a, 145/127; cf. 96/85). Derrida points out that, in each of these binary oppositions, the two terms do not stand on an

148). Barad's approach in part converges with my reading of Bohr in the present study, and deserves close examination to be conducted on another occasion.

[6] Somewhat similarly to the case of 'complementarity,' the term 'deconstruction' is also often used in a double sense: On the one hand, it refers to a specific kind of critical operation on metaphysical ideas, and on the other, to an overall philosophical project revolving around this deconstructive operation.

equal footing, but in a definite "order of subordination" (1972b, 392/329; cf. 1972c, 56f./41). That is, the first term is valued as superior, originary, and independent, while the second is devalued as inferior to, derivative from, and dependent on the former. In this way, the binary oppositions characteristic of the Western philosophical tradition are "hierarchical," where one term of each pair is privileged over the other.

According to Derrida, the above hierarchical binary oppositions are closely connected with each other to form a system that he calls "metaphysics." In order to explicate this, he introduces the two key terms "logocentrism" and "phonocentrism" to denote two major conceptual strands of metaphysics (1967b, 23/12; cf. 11/3). The term 'logocentrism' refers to the philosophical orientation toward an order of meaning conceived as existing in itself.[7] It implies the notion of a "transcendental signified," which "in and of itself, in its essence, would refer to no signifier, would exceed the chain of signs, and would no longer itself function as a signifier" (1972c, 30/19f.; cf. 41/29; 1967b, 33/20, 71f./49).[8] 'Phonocentrism,' on the other hand, means the privileging of *phōnē* or the voice, or, put differently, the hierarchical opposition of speech/writing.[9] What is crucial to Derrida is that these two components of metaphysics, logo- and phonocentrism, are inextricably bound up with each other in the following manner. The privileging of the voice and speech rests on the notion that, in spoken language, what is meant by the speaker is understood immediately and in its full presence within the given context. This is because the voice appears to serve as "the signifying substance given to consciousness as that which is most intimately tied to the thought of the signified concept" (1972c, 32/22). The spoken sign seems even to "fade away" as a signifier and become "transparent," thus "allow[ing] the concept to present itself as what it is, referring to nothing other than its presence" (1967c, 86/77; 1972c, 33/22).[10] This indicates "an essential tie between *logos* and *phōnē*" (1967c, 14/15; cf. 1967a, 293/196; 1967b, 45/29), which has been supposed to ensure the "immediate presence" of meaning as "absolute proximity to oneself" (1967c, 86/77, 65/58; cf. 111/99). Written language, in contrast, has traditionally been seen as "that which threatens presence," specifically the presence of signified meaning (1967a, 293/197). For the written sign is capable of being repeatedly read "in the absence of and beyond the presence of the empirically determined subject" and, by the same token, "carries with it a force of breaking with its context" (1972b, 377/317; cf. 372/313, 375f./315f.). For this reason, writing has

---

[7] Since the Greek word *logos* has a range of meanings such as word, speech, thought, and reason, Derrida's term 'logocentrism' also tends to vary in meaning depending on the context, at times coming close to or overlapping with the term 'phonocentrism' (see 1967b, 11/3).

[8] Derrida also characterizes the exigency of the transcendental signified as the demand "that the difference between signified and signifier [be] absolute and irreducible" (1967b, 33/20).

[9] See, for example, Culler (1982, 92f.). It is worth recalling that, in Section 5.1, we saw how Gadamer privileges speech as living language over writing as "the dead trace of meaning." For a deconstructive critique of Gadamer's hermeneutics with a focus on this phonocentrism, Takahashi (1992, 235–65).

[10] See also Derrida (1967b, 33/20; 1967c, 9/10, 16/16).

been debased as a mere representation of speech or even excluded from the concept of language (see 1967b, 12/3; 1972c, 36/24f.).[11]

In Derrida's view, this intrinsic connection between phonocentrism and logocentrism also indicates a pivotal role played by the notion of "presence." As has been seen above, in metaphysics, speech is privileged over writing because the former supposedly makes possible the presence of meaning in the presence of the speaking subject as well as of the original context, whereas the latter, writing, implies a break with this "set of presences" (1972b, 377/317). We can readily see that this metaphysical conception rests on the privileging of presence over absence – or, in certain contexts, of presence over representation or of the present over the non-present – which may also be characterized as "the determination of being in general as presence" (1967c, 83/74; cf. 59/53; 1967a, 411/279).[12] Since the link between *logos* and *phōnē* thus proves to be based on the supposed primacy of presence, Derrida designates the overall framework of the metaphysical tradition as "the metaphysics of presence" (1967c, 57/51).

Derrida's critical engagement with this metaphysics of presence proceeds through a close textual analysis of the work of a series of Western philosophers and thinkers such as Plato, Aristotle, Rousseau, Hegel, Saussure, Husserl, and so forth (see Derrida 1972c, 15/7, 33/22). His critical reading of Saussurean semiology is of particular importance here. On the one hand, Derrida points favorably to the aspects of Ferdinand de Saussure's thought that tend to go beyond the metaphysical tradition: the idea of the inseparability of "signified (*signifié*)" and "signifier (*signifiant*)" as well as of the "differential and formal characteristics" of linguistic signs (Derrida 1972c, 28/18). On the other, Derrida contends that, by maintaining the rigorous distinction between signified and signifier, Saussure "leaves open the possibility of thinking a concept signified in and of itself," that is, a transcendental signified (1972c, 29f./19). Correlatively, he continues, Saussure "privilege[s] speech, everything that links the sign to *phōnē*," over writing (1972c, 31/21). That is, supposing the "natural link" (Saussure 1916, 46/26) between thought and voice, meaning and sound, he regards writing "as a phenomenon of exterior representation, both useless and dangerous," thus excluding it from the system of language (*langue*) (Derrida 1972c, 31f./21, 36/24f.; cf. 49/36).

Similarly but more extensively, Derrida also develops a critical reading of Husserlian phenomenology – an analysis which played a pivotal role in the

---

[11] In this connection, Derrida maintains that nonphonetic writing developed in science, especially in the modern mathematical sciences, tends to deviate from the metaphysical tradition. As he puts it, "the practice of science has constantly challenged the imperialism of the logos, by invoking, for example, from the beginning and ever increasingly, nonphonetic writing" (1967b, 11/3, trans. mod.; cf. 112–21/76–81). Conversely speaking, "the resistance to logical-mathematical notation has always been the signature of logocentrism and phonologism" (1972c, 46/34). See Plotnitsky (1994, 62).

[12] According to Derrida, this notion of presence has historically taken various forms such as "*eidos, archē, telos, energeia, ousia* (essence, existence, substance, subject), *alētheia*, transcendentality, consciouness, God, man, and so forth." Since the late nineteenth century, however, these metaphysical notions have also been variously put in question, specifically by Nietzsche and Freud, and "more radically" by Heidegger (Derrida 1967a, 411f./280).

formation and elaboration of his project of deconstruction itself. In his account, Edmund Husserl's philosophy, while undermining the metaphysical tradition in considering "the themes of temporalization, and of the relationship to the alter ego" (1967a, 196f./134; cf. 178/121; 1967c, 40/37, 58/52), is still dominated by the metaphysics of presence "in its most modern, critical, and vigilant form" (1972c, 13/5; cf. 1967a, 249/166; 1967b, 72/49). Specifically, in his early work *Logische Untersuchungen* [*Logical Investigations*], distinguishing between two kinds of signs, "expression (*Ausdruck*)" and "indication (*Anzeichen*),"[13] Husserl emphasizes that expressions operate purely – without intervention by indications – in "monological speech." For while, in communication, "meaning is always interwoven with [...] an indicative function," in monologue words perform no indicative, but solely an expressive function, since their meaning is experienced "at the same moment" by the subject (Husserl 1913, 24, 36f.).[14] In Derrida's account, Husserl thus privileges not simply speech, but specifically monological speech – which Derrida characterizes as "hearing oneself speak (*s'entendre parler*)" (1967c, 88/78; cf. 29/27f., 83/74; 1967a, 248/166) – on the ground that, in monologue, the signified meaning seems to be fully and immediately present to consciousness (see 1972c, 43f./31f.). In this way, logo- and phonocentrism are inextricably linked in Husserl's philosophy, too, as a specific form of the metaphysics of presence.

Let us now look closely at the way in which Derrida seeks to "deconstruct" the metaphysics of presence as outlined so far. Derrida argues that the above privilege of speech over writing, bound up with the central notion of presence, is "only apparent" (1967c, 86/77). Here, without entering into the details of his reading of individual thinkers, I will trace the major line of his argument. According to Derrida, the character of writing as repeatable and separable from the context, which has traditionally been contrasted with speech's privileged access to presence, proves in reality to extend beyond writing in the ordinary sense. Indeed, any linguistic sign, whether spoken or written, can serve as such precisely because and insofar as it "may be indefinitely repeated as the same" (1967c, 8/9). Derrida amplifies this in *La voix et le phénomène*:

> When in fact I *effectively* use words [...], I must from the outset operate (within) a structure of repetition whose basic element can only be representative. A sign is never an event, if by event we mean an irreplaceable and irreversible empirical particular. A sign which would take place but "once" would not be a sign [...]. A signifier (in general) must be formally recognizable in spite of, and through, the diversity of empirical characteristics which may modify it. It must remain the same, and be able to be repeated as such [...] (Derrida 1967c, 55/50; cf. 88/78).

In other words, the identity of a linguistic sign depends entirely on, and "is constituted [...] by the possibility of being repeated," repeated even, and especially, in the absence of the subject and in separation from the context (1972b, 378/318;

[13] By "expressions" Husserl means meaningful signs, and by "indications" the signs that "express nothing" and are devoid of meaning (1913, 23). See Derrida (1967c, 17/17).

[14] See Derrida (1967c, 20/20, 54/49) and also Evans (1991, 33, 74).

cf. 1967c, 58/52, 60/54, 84/75). Repeatability or "iterability," traditionally held to be characteristic of writing, thus proves to be constitutive of language in general (1972b, 378/318; cf. 1967c, 64/57).

For this reason, Derrida states that all language is "first, in a sense [...], writing" (1967b, 55/37; cf. 63/43). To be sure, he is in no way attempting here to "rehabilitat[e] writing in the narrow sense," or simply to "revers[e] the order of dependence" (1967b, 82/56; cf. 1972c, 22/12). Rather, he seeks to introduce "a new concept of writing," a concept of "generalized writing" or "arche-writing (*archi-écriture*)" to which also "oral language already belongs" (1972c, 37/26; 1967b, 81/55, 83/57).[15] In Derrida's account, this generalized concept of writing – or, more or less equivalently, the notion of iterability – may be extended to "the sign in general" (1972b, 12/12), and further to any system of "marks" and even "all 'experience' in general" (1972b, 378/318; cf. 377/316f.).

Further, as has already been suggested, the repetition of a sign or mark, while constitutive of its ideal identity, is not a simple reproduction of the identical. Rather, emphasizes Derrida, it "always alters [...] that which it seems to reproduce" (1990, 82/40; cf. 120/62). This is why he prefers the term 'iteration' to repetition, and 'iterability' to repeatability, noting that the prefix *iter* etymologically means 'other' (1972b, 375/315). As Derrida puts it,

> [...] the structure of iteration implies both identity and difference. Iteration in its "purest" form – and it is always impure – contains in itself the discrepancy of a difference that constitutes it as iteration. The iterability of an element divides its own identity a priori (Derrida 1990, 105/53; cf. 135/71; 1967c, 15/15).

In other words, while iterability enables any element to remain itself, to be identical to itself, it paradoxically "ruins [...] the identity" thus rendered possible (1990, 144f./76). It further follows that "repetition does not *happen* to" the initial occurrence, but "its possibility is already there" from the beginning (1967a, 301/202; cf. 314/211). That is, instead of repeating an already constituted entity, iterability "divides the point of departure of the first time" (1967a, 301/213).

Closely linked with the above ideas of arche-writing and iterability, Derrida introduces yet another key term, "*différance*." As a modification of the French verb *différer*, which means both 'to differ' and 'to defer,' his neologism *différance* correspondingly has the following two meanings, or rather two major clusters of meanings (see 1972c, 16f./8).[16] On the one hand, *différance* signifies the "production of differences" or the "process of differentiation" (1972c, 39/28, 60/101; cf. 1972b, 12/11).[17] This term accordingly overlaps with that of iterability just considered – iterability which alters any sign or mark in its repetition and thus produces

[15] For parallel passages, see Derrida (1967a, 294/197; 1967b, 65/44, 68/46, 74/51).

[16] See, for example, Gasché (1986, 194ff.) and also Plotnitsky (1994, 39).

[17] Derrida often tends, however, to avoid using the term "differentiation," insofar as it would suggest "an organic, original, and homogeneous unity that eventually would come to be divided" (1972b, 14/13).

differences "within each individual 'element'" (1990, 105/53).[18] The notion of *différance* also concerns, however, the production of differences *between* the elements. In other words, it is constitutive of the "diacriticity," namely the differential character of signs which Saussurean linguistics has shown to be "the condition for any signification and any structure" (1972c, 17/9). In this sense, *différance* may be defined as "the movement according to which language, or any code, any system of referral in general, is constituted 'historically' as a weave of differences" (1972b, 12/12; cf. 1972c, 38f./27).[19] This movement of *différance* covers also the production of *différends* or of "all the oppositional concepts that mark our language," including the hierarchical binary opposites characterizing metaphysics (1972c, 17/9; cf. 1972b, 8/8).

On the other hand, continues Derrida, *différance* also means deferral, delay, and "temporization" (1972b, 19/18; cf. 1972c, 17/8). More precisely, it signifies the circumstance that, by means of detour and postponement, "the relationship to the present, the reference to a present reality [...] are always deferred" (1972c, 40/29; cf. 1972b, 10/10, 15/14). Here we can perceive another crucial link with his above concept of iterability. That is, the idea of iterability – the possibility of the indefinite iteration of signs – implies that signs defer or put off indefinitely access to that of which they are signs.[20] Moreover, this indefinite delay is also entailed by what Derrida indicates as the breakdown of the notion of the transcendental signified: In his account, there is no signified meaning outside the chain of signs, but rather every signified itself "signifies again," that is to say, "is also in the position of a signifier" (1967a, 42/25; 1972c, 30/20; cf. 1967a, 302/203).[21] For these reasons, access to the signified meaning as such is indefinitely deferred – a movement which is nothing other than *différance* as deferral.

The above two sets of meanings taken together, *différance* may be summarily designated as the "originary" movement or play of differing and deferring (1967a, 302/203; cf. 1967b, 88/60). We can see that many basic strands of Derrida's thought as reviewed so far join together in this key term. Further, he critically characterizes hierarchical binary oppositions in terms of *différance* with the above double

---

[18] This meaning of *différance* is not explicitly discussed in Derrida's article "La différance" (in 1972b, 1–29/1–27) or in his concise account of the term in *Positions*. In the latter, he mentions the following four distinct features of *différance*: the movement of deferring, "the common root of all [...] oppositional concepts," the production of diacriticity, and the unfolding of the "ontico-ontological difference" (1972c, 17–19/8–10). In his later work *Limited Inc.*, however, Derrida speaks of "repetition 'as' *différance*," which includes the sense of altering each element in its repetition (1990, 107/54).

[19] The word 'historically' is used here in reference to Saussure's remark: "historically, the fact of speech always comes first [in relation to the language]" (Saussure 1916, 37/19, trans. mod.). Derrida employs the term 'history' with due caution, however, concerning its possible teleological connotation of "a final repression of difference" (1972b, 12/11).

[20] For an analysis of this point and its similarities to Donald Davidson's ideas, see Wheeler III (2000, 28ff.).

[21] For parallel passages, see Derrida (1967a, 311/209, 412/281; 1967b, 16/7, 36/23, 108/73; 1972c, 38/26).

meaning: In each of the binary oppositions, one of the terms, the devalued term, "appear[s] as the *différance* of the other, as the other different and deferred" (1972b, 18/17; cf. 9/9; 1972c, 41/29). In the case of speech/writing, for example, writing appears as speech *différé-différant*, because writing not only differs from speech, but also defers the latter's supposedly immediate access to meaning. Whereas, from the metaphysical point of view, writing is *merely* a differing-deferring modification of speech, Derrida subverts this hierarchical order precisely by virtue of the "originary" character of *différance*. At the same time, however, he exercises special caution in using the term 'originary':

> To say that *différance* is originary is simultaneously to erase the myth of a present origin. Which is why "originary" must be understood as having been *crossed out* [...] (Derrida 1967a, 302/203; see Plotnitsky 1994, 44).

In other words, *différance* is the "non-full, non-simple, structured and differentiating origin," so that "the name 'origin' no longer suits it" (1972b, 12/11).

According to Derrida, this originary and yet non-originary character of *différance* prevents any word or concept – indeed, even the term *différance* itself – from "coming to summarize and to govern" uniformly the textual play of differences (1972c, 23/14). Rather, the term *différance* "lends itself to a certain number of nonsynonymous substitutions," which includes not only arche-writing and iterability as already seen, but also the "trace," "spacing," the "supplement," the *pharmakon*, and so on (1972b, 13/12; cf. 28/26; 1972c, 16/8).[22] Among these terms, here I will comment only on the trace and the supplement (see 1967b, 92/62, 95/65).[23] The movement of *différance* as sketched above implies that it is impossible for a simple (linguistic or other) element to "be *present* in and of itself, referring only to itself." As already suggested by Saussure and radically reconceived by Derrida, any element can function as a sign or mark only by referring to other elements. Derrida renders this circumstance in terms of the "trace": Each element is "constituted on the basis of the trace within it of the other elements of the chain or system" (1972c, 38/26; cf. 1967c, 95/85). The trace thus designates the general structure required for "the relationship with the other" or the relation to "radical alterity" as constitutive of what appears to be present (1967b, 69/47; 1972b, 21/21; cf. 25/24).[24] Since this

---

[22] In his deconstructive reading of Plato's *Phaedrus*, Derrida focuses on the double meaning of the Greek *pharmakon* – cure and poison – the pivotal term Plato employs in his criticism of writing (Derrida 1972, 69ff./61ff.).

[23] In his systematic interpretation of Derridean deconstruction, Rodolph Gasché discusses what he terms the "infrastructural chain" which consists of arche-trace, *différance*, supplementarity, iterability, and re-mark (1986, 185–224).

[24] It is noteworthy that Derrida reconceives "the reality of the external world" as "alterity" in the sense of radical alterity (1967a, 182/124). He also connects the notion of radical alterity with that of "matter": "if, and in the extent to which, *matter* in this general economy designates [...] radical alterity (I will specify: in relation to philosophical oppositions), then what I write can be considered 'materialist'" (1972c, 87/64). See Plotnitsky (1994, 57f.). Derrida's concept of radical alterity is closely related to, yet distinct from, Emmanuel Levinas's notion of the "absolute other." For Derrida's critique of the latter, see especially 1967a, 125ff./84ff.

trace is never present as such, it follows that "[n]othing [. . .] is anywhere ever simply present or absent," but that "[t]here are only, everywhere, differences and traces of traces" (1972c, 38/26; cf. 1972b, 24/23, 76f./66).

Further, similarly to *différance*, the notion of the trace also has a temporal dimension. In Derrida's view, for any element to serve as a sign, it must not only be related to other elements on the synchronic plane, but also – as his discussion of iterability shows – must be repeatable in the non-present, in the past as well as the future. To put it in terms of the trace, each "so-called 'present' element" must bear within itself the traces of past and future elements. The trace thus "constitut[es] what is called the present by means of [the] very relation to what is not" (1972b, 13/13). Unlike Husserl's concepts of retention and protention, the trace cannot be thought "on the basis of the present, or of the presence of the present." Rather it should be conceived as the structure of referral to "a 'past' that has never been present" and to a future that will never be "in the form of presence" (1972b, 21/21; cf. 1967b, 125/84). Derrida accordingly characterizes the trace – in the temporal and the above non-temporal aspects taken together – the "originary trace or arche-trace" (1967b, 90/61). This designation is again paradoxical, however. For to say that the trace is "the absolute origin of sense in general" amounts to saying that "there is no absolute origin of sense in general" (1967b, 95/65; cf. 1967a, 302/303; 1967c, 76/68; 1972b, 12/11).

While the trace thus focuses on the structure of referral to the other, the supplement, another major Derridean term, designates the structural need of *adding* an other (1967b, 215/150; cf. 1967c, 98/88).[25] This term has been borrowed from Jean-Jacques Rousseau, who, in a phono- and logocentric manner, criticized writing as a mere "supplement" to speech (see 1967b, 207/144). Derrida undermines Rousseau's view, however, by arguing that writing can be added to speech because speech is not an original plenitude, but, just as writing, has a lack – the lack of fully present meaning – which enables writing to supplement it. He thus critically displaces the concept of the supplement in such a way that, while seemingly adding itself to plenitude, the supplement proves to fill an absence of plenitude, to "replace" or "insinuate itself *in-the-place-of*" an absent origin (1967b, 208/145; cf. 1967a, 314/212, 423f./289f.; 1972a, 193/167). As Derrida puts it, the "strange structure of the supplement" is such that "by delayed reaction, a possibility produces that to which it is said to be added on" (1967c, 99/89). In this way, what is called the origin is always preceded by the supplement, which belatedly reconstitutes the origin by taking the place of its absence (see 1967a, 332/224; 1967b, 308/215).

We have so far seen in outline the way in which Derrida seeks to undo and dismantle the metaphysics of presence – or rather to show how it dismantles itself – and correlatively develops his new ideas of arche-writing, iterability, *différance*, and so forth. According to Derrida, this critical operation on the metaphysical system, which he terms "deconstruction," consists of the following two phases or moments. First, given the metaphysics of presence as a system of hierarchical binary oppositions of speech/writing, presence/absence, self/other, origin/repetition, identity/

[25] See Gasché (1986, 205f.).

difference, and so on, deconstruction reverses or "overturn[s]" the "order of subordination," making the privileged term (speech, for example) dependent on the devalued one (writing). This "phase of overturning" is necessary because of "the conflictual and subordinating structure of opposition." Without going through this phase, "one might proceed too quickly to a *neutralization* that *in practice* would leave the previous field untouched [. . .], thereby preventing any means of *intervening* in the field effectively" (1972c, 57/41; cf. 1967a, 403/274). Although rather tentatively called a "phase," this overturning is not a "chronological phase" that could be left behind in a certain stage, but is structurally necessary as an "interminable analysis" (1972c, 57/41f.). Second, however, Derrida continues, deconstruction does not restrict itself to the above inversion of the hierarchical order. Rather, instead of preserving the reversed hierarchy, it carries out a "general displacement of the system" (1972b, 392/329), producing "new concepts and new models, an *economy* evading [the] system of metaphysical oppositions" (1967a, 34/19, trans. mod.; cf. 1967b, 82/56; 1972c, 57/42).[26] *Différance*, arche-writing, and other associated structures as seen above are such new notions which serve to "disorganiz[e] the entire inherited order and invad[e] the entire field" (1972c, 58/42). It is by means of this "double gesture" of inversion and displacement that deconstruction seeks to intervene critically and effectively in the metaphysical system (1972b, 392/329).

This also enables us to see, in particular, how deconstruction differs from a clear-cut "rupture" or an absolute "transgression" (1972c, 35/24). According to Derrida, since "[t]here is no sure opposition between outside and inside," there is no transgression if one understands by this term "a pure and simple landing into a beyond of metaphysics" (1972c, 2/12; cf. 1967a, 403f./274). Rather than absolutely breaking with metaphysics, deconstruction is in a way "caught within the metaphysical closure (*clôture*)" (1967b, 148/99; cf. 14/4; 1967a, 296/198f.; 1972c, 15/6). That is, even, and especially, in the practice of deconstruction, one cannot avoid using notions belonging to the metaphysical tradition – such as "sign, history, truth, and so on" – and, in this sense, one remains dependent on and enclosed in what is deconstructed (1967a, 421/288; cf. 1967b, 89/60; 1972c, 19/10). In other words, one must "borrow from a heritage the resources necessary for the deconstruction of that heritage itself" (1967a, 414/282; cf. 412/280f.). Derrida's term 'closure' or "the closure of metaphysics" (1967c, 57/52; cf. 1967a, 399/272; 1972c, 21/12, 77/56) thus designates the paradoxical relation "between belonging and the opening," that is, "the necessity of lodging oneself within traditional conceptuality in order to destroy it" (1967a, 163/110, 165/111; cf. 34/20, 240/162).[27] It is owing to this closure that deconstruction is pursued as an interminable project of undoing and transforming metaphysics by turning its conceptual system against itself (see 1972c, 35/24; Plotnitsky 1994, 225f.).[28]

[26] See Plotnitsky (1994, 61).

[27] Simon Critchley pursues Derrida's motif of the closure of metaphysics in relation to the thematic of Levinasian ethics (Critchley 1992, esp. 59–106).

[28] For this reason, Derrida emphasizes the difference between "closure" and "end," maintaining that "[w]hat is held within the demarcated closure may continue indefinitely" (1972c, 23/13; cf. 1967b, 14/4).

In this section, we have traced Derrida's main lines of argument in his early work of deconstruction, notwithstanding its intrinsic tendency to elude any attempt at summarization. To be sure, similarly to hermeneutic philosophy as reviewed in Chapter 5,[29] Derrida's work largely proceeds in the closest association with issues developed in the humanities, not in the natural sciences.[30] Yet his thematic of deconstruction itself, with its basic ideas of iterability, *différance*, the trace, and so on, appears to be so general and far-reaching as to pertain to natural science no less than to any other fields. It is therefore no wonder that the question is posed whether Derridean deconstruction has points of contact with Bohr's idea of complementarity, which in turn is marked by its philosophical implications reaching far beyond the confines of quantum physics. In the next section, let us accordingly go on to survey Plotnitsky's analysis of the possible conceptual links between the two thinkers' ideas.

6.2 Plotnitsky on Complementarity and Deconstruction

Needless to say, it is far from commonplace to associate Bohr's complementarity and Derridean deconstruction, which might seem to be disparate sets of ideas developed in entirely unconnected scholarly fields. Some researchers have suggested, however, certain conceptual affinities between these two approaches of thought, particularly in the way they each question and seek to undo traditional scientific-philosophical ideas and theories.[31] In fact, in his 1994 book *Complementarity: Anti-epistemology after Bohr and Derrida*[32] and other related works, Arkady

[29] In the present study, I do not enter thematically into the relation between hermeneutics and deconstruction. For an attempt to radicalize hermeneutic philosophy through interactions between Heidegger's and Derrida's thought, see Caputo (1987). Gadamer and Derrida directly exchanged views at a Paris symposium in 1981, and their texts, together with commentators' accounts, have been compiled into Forget (1984) as well as Michelfelder and Palmer (1989). Derrida's farewell speech following Gadamer's death is included in Derrida and Gadamer (2004). For a systematic inquiry into similarities and differences between Gadamerian hermeneutics and Derridean deconstruction, see Bertram (2002). Further, on the relation between Ricoeur's and Derrida's philosophical approaches, see Lawlor (1992).

[30] Admittedly, in his earliest period, Derrida addressed philosophical problems of mathematics in his "Introduction" to Edmund Husserl's *L'origine de la géométrie* (Derrida 1962, 3–171/23–153).

[31] Almost simultaneously with Plotnitsky's *Complementarity*, John Honner also pointed to similarities between Bohr's and Derrida's ideas. In a brief article, criticizing many philosophical studies of Bohr's thought for trying to catch it "in the wrong kind of net," Honner emphasizes that complementarity is concerned with the circumstance that "the practical univocity of our ordinary descriptive concepts breaks down" in a way that suggests a linkage with Derridean deconstruction. John Honner, "Description and Deconstruction: Niels Bohr and Modern Philosophy," in Faye and Folse (1994, 141–53, on 144f. and 148). Notwithstanding this innovative insight, however, his actual account in the paper seems to be no more than a tentative sketch and also limited by a partly inadequate characterization of Derrida's thought.

[32] Plotnitsky defines the term "anti-epistemology" as "the general possibility of a dislocation, or [...] deconstruction of classical or metaphysical theories – epistemologies, ontologies, phenomenologies, or, to return to Derrida's more encompassing terms, forms of ontotheology, logocentrism, and the metaphysics of presence" (1994, 10), so that the term is more or less

Plotnitsky has discussed significant intersections of Bohr's complementarity with what is called "general economy," above all with the latter's contemporary radicalized form, deconstruction (1994, 1). Even though relatively little known among historians and philosophers of science, Plotnitsky's analysis appears to deserve full attention across the disciplinary boundaries.[33]

Tracing the texts of Bohr's Como lecture and subsequent works, Plotnitsky starts by discussing what he sees as Bohr's two major forms of complementarity in quantum theory: "the complementarity of space-time coordination and the claim of causality" – or, as Plotnitsky puts it for brevity, that of "coordination and causality"[34] – and "wave-particle complementarity" (1994, 68f.). Although the former form of complementarity appears as the most important for Bohr's overall project, neither of the two may be regarded as the cause or ground of the other or any other forms of complementarity (see 1994, 69).[35] One can see, in particular, that the complementarity of coordination and causality "dissociat[es] that which is always united" in classical physics, whereas wave-particle complementarity "combin[es] that which is always dissociated" there (1994, 5f.; cf. 122). In Plotnitsky's account, Bohr's complementarity thus undoes both "the classical, unequivocal unifications and classical, unequivocal dissociations" of descriptive features (1994, 6; cf. 22, 132).

Bohr's idea of complementarity is not, however, restricted to those relations which he explicitly designates by the term (see 1994, 121). Rather, Plotnitsky argues, it further implies complex relations between "continuity and discontinuity," "chance and necessity," and other associated pairs of notions.[36] Taking a step beyond Bohr's terminology, Plotnitsky proposes that these conceptual pairs be also called complementary (1994, 123, 125; cf. 217, 232). Further, as Bohr

---

equivalent to 'general economy.' In his more recent work, however, he has distanced himself from the use of the term 'anti-epistemology' in favor of such designations as "nonclassical thought" and "nonclassical epistemology" (2002, 1; 2006, 143ff.). See also Smith and Plotnitsky (1997).

[33] In his more recent texts including *The Knowable and the Unknowable* (2002), Plotnitsky tends to connect less directly Bohr's complementarity and Derridean deconstruction. The focus of this section is not so much on these works as on *Complementarity* and other writings published in the 1990s.

[34] In *The Knowable and the Unknowable*, however, Plotnitsky notes that "the complementarity of coordination and causality was to disappear rather quickly from Bohr's writings," with 'the claim of causality' replaced by a phrase such as 'the application of the laws of conservation of energy and momentum' (2002, 59).

[35] This characterization by Plotnitsky still seems to overestimate the importance of wave-particle complementarity. As we saw in Section 2.1 and subsequently, Bohr's term 'complementarity' with regard to quantum theory refers primarily to the relation between space-time coordination and the claim of causality (or dynamical conservation), and not to the wave-particle relation. Plotnitsky himself later ceased to emphasize wave-particle complementarity, noting that this form of complementarity "was never especially favored by Bohr" (2006, 12; cf. 18, 125).

[36] We can see not only that the wave picture is continuous and the particle picture discontinuous (see 1993b, 86), but also, at a 'metalevel,' that the above-mentioned unification and dissociation of descriptive features correspond, respectively, to continuity and discontinuity.

himself expands his scope "well beyond the domain of quantum physics" (1994, 72; cf. 1993b, 55, 68, 79ff.), it becomes more evident that complementarity – as a theoretical "framework" or "matrix" (1994, 68f.; cf. 75) – concerns such "general concepts and conceptual structures [...] as subject and object, interiority and exteriority, analysis and synthesis [...] and so forth" (1994, 40; cf. 22, 121). Bohr, so to speak, "complementariz[es]" these conceptual pairs, which have commonly formed binary oppositions in the philosophical tradition (1994, 232; cf. 132, 138). According to Plotnitsky, this extended notion of complementarity designates generally the diverse configurations that are "heterogeneously interactive and interactively heterogeneous" (1994, 12; cf. 24, 73; 1993b, 47).

Here one might wonder, however, whether this extension of complementarity accords with Bohr's own views. To consider this question, we should refer to Plotnitsky's own characterization of his inquiry: His work on Bohr's complementarity is not simply of expository nature, but contains an attempt to develop complementarity "more comprehensively than Bohr does" (1994, 73). As Plotnitsky notes, Bohr defines the term 'complementarity' to mean "both mutual exclusivity and completeness of description" (1994, 5). Plotnitsky maintains, however, that, in fact, "complementary constituents are not always mutually exclusive," if complementarity is viewed as a "general matrix" (1994, 75). This, together with other considerations, leads him to extend and generalize the idea of complementarity in such a way as to cover "diverse – and at times conflicting or mutually incompatible (particularly from the classical perspective) – configurations, double or multiple, operative within the same framework, but without lending themselves to a full synthesis" (1994, 73).[37] This redefinition of complementarity also indicates another aspect of Plotnitsky's conceptual extension: Complementarity is no longer limited to a two-term relation. In his more explicit account, "complementary relationships may be extended into triple or more multiple configurations, some of which cannot be controlled by dualities" (1994, 75).

Bohr's idea of complementarity, thus interpreted and recomprehended, appears to have far-reaching philosophical implications, suggesting possible intersections with a wide range of philosophical discussions. Plotnitsky considers, however, many existing philosophical interpretations of Bohr's thought to be highly problematic. The major interpretive studies so far have largely revolved around the conventional opposition of realism and anti-realism, and Plotnitsky first critically assesses the realist reading, as exemplified by Henry Folse's analysis. As we reviewed in Section 3.2, in reference to Bohr's account of complementarity as a relation between two different phenomena with regard to a single atomic "object," Folse claims that this conceptual scheme postulates the reality of the object "existing independently of observational interactions" (1985, 238, 151). In Plotnitsky's view, however, neither Bohr's use of the term 'object' or 'reality' nor his overall framework of

[37] For parallel passages, see Plotnitsky (1993a, 5, 9; 1993b, ixv; 1994, 10, 24).

complementarity allows such a realist reading.[38] Drawing, in particular, on Bohr's early assertion that the notion of independent reality "can neither be ascribed to the phenomena nor to the agencies of observation" as well as his advocacy, made in his response to EPR, of "a radical revision of our attitude towards the problem of physical reality" (PWNB, 4:75; see Plotnitsky 1994, 100f., 161), Plotnitsky argues that "[i]n Bohr's matrix, no independent physical reality or object can exist" (1994, 117; cf. 58; 1993b, 32, 35). It is, in Plotnitsky's account, also along this line of thought that Bohr later criticized expressions such as "disturbing of phenomena by observation." That is, he considered such a phrase to be inappropriate because it would suggest the notion of an "undisturbed" state of reality prior to and independent of observation.[39] One must, continues Plotnitsky, also be cautious about Bohr's usage of the term "objectivity" (1994, 114f.; cf. 82). When defending the objective character of quantum-theoretic knowledge, Bohr does not mean the objectivity of an object existing independently of the knowing subject, but rather "refers to the conditions of possibility of unambiguous communication of the experimental results" (1994, 116; cf. 102, 108, 114).[40]

This might appear to favor the opposite camp of commentators who characterize Bohr's thought as "idealism," "positivism," or other forms of anti-realism.[41] According to Plotnitsky, however, this type of interpretation is no more warranted than the above realist type of reading.[42] While rejecting the classical notion of independent reality, Bohr also distances himself from the view "that what is not observed, or even what cannot be observed, does not exist" (1994, 101f.). For Bohr, what is called reality cannot be reduced to observability, nor is it simply a conceptual creation or construction by the knowing subject. Rather, there is an aspect of reality – designated by Plotnitsky as "material efficacity" – that "affects and constrains all observation, measurement, interpretation, and theory" and yet is not fully

[38] Plotnitsky rejects also Dugald Murdoch's interpretation, according to which Bohr holds that "successful observation or measurement reveals the objective, preexisting value of an observable" (Murdoch 1987, 107; see Plotnitsky 1994, 80).

[39] While this view is shared by the majority of commentators, as reviewed in Section 3.3, I presented an alternative interpretation in Section 4.2.

[40] Plotnitsky is nevertheless critical of Bohr's very use of the term 'objective,' which, in his view, tends to obscure the latter's radical break with any traditional notion of objectivity (1994, 116f.). It seems to me, however, that what would be problematic from a deconstructive point of view is not merely Bohr's *term* 'objectivity,' but rather his conception of it in certain contexts. That is, the notion of objectivity as unambiguous communicability (or univocal intersubjective validity mediated by linguistic communication) is not radical or deconstructive, but rather quite common and perhaps even standard in twentieth-century philosophy in such a way that it could be readily *subject* to deconstructive critique. As discussed in Chapter 4, this conception of objectivity, which comes to the foreground in Bohr's later work, should be distinguished from the more radical moments and aspects of his thought.

[41] Plotnitsky characterizes "positivism," in particular, as an "uncritical obverse" of "the metaphysics of presence" (1994, 4). He also critically comments on Jan Faye's account of Bohr's thought as "objective anti-realism," while agreeing with some of his points against realist readings (1994, 81f.).

[42] See Plotnitsky (1993b, 32, 35; 1994, 85, 101, 115f., 120, 172).

accessible to observation or theoretical conceptualization (1994, 94, 58). It is in this sense that we can grasp Bohr's simultaneous criticism of the expressions "disturbing of phenomena by observation" *and* "creating physical attributes to atomic objects by measurements" (see 1994, 115).

According to Plotnitsky, we can now see that, while continuing to use classical terms such as 'reality,' 'phenomenon,' and 'object,' Bohr significantly displaces their meanings (see 1994, 82, 101, 113). At this point, Plotnitsky gives his reading an important terminological twist: Bohr's displaced notion of reality, in particular, may be better designated as "alterity" in a Derridean sense. In his account:

> One might do well to abandon the term 'reality' altogether, provided that one takes precautions against positivist, idealist, phenomenologist, or transcendentalist interpretations that would reverse this concept without sufficiently displacing the metaphysical base grounding it. [...] Rhetorically or strategically, alterity is a much better term [...] (Plotnitsky 1994, 108).[43]

By this alterity Plotnitsky does not mean "absolute alterity," which would be akin to the Kantian thing-in-itself (see 1994, 22, 79, 259) or conform to "negative ontotheology" (1994, 119; cf. 35, 46, 53). Bohr's thought does not refer to the absolutely 'other,' but rather operates in a "complementary" or "reciprocal" relation between self and other, inside and outside, or subject and object. It nevertheless concerns itself with the irreducibly and radically other, an other which is in no way secondary or derivative to the self. According to Plotnitsky, this conception of "radical alterity," which is at work in Bohr's overall framework, has been missed by most prior readings of Bohr including both realist and anti-realist ones (1994, 108; cf. 115; 1993b, 311). In other words, "most, perhaps all analyses of Bohr so far" have ignored such a novel philosophical dimension of complementarity, thus reenclosing it within "classical epistemologies" (1994, 82, 79; cf. 77, 113, 118). How, then, more specifically, does Bohr's thought revolve around the above notion of radical alterity, and how can it be further associated with Derridean deconstruction?

As we have seen, one of Bohr's basic points is that any observation of atomic phenomena involves an unavoidable and uncontrollable "interaction between the object and the instrument of observation" (PWNB, 1:93). This implies that observation carries with it an "inevitable loss of knowledge," as is exemplified by the case in which the measurement of the position of an atomic object is accompanied by a loss of knowledge of its momentum (PWNB, 4:78; see Plotnitsky 1994, 100). As Plotnitsky notes, this loss of knowledge is not a loss of something – in the above case, a definite value of momentum – that was originally present. Rather, it is a radical and 'originary' loss which is analogous to the "loss of meaning" in Bataillean general economy (1994, 1),[44] and particularly to its Derridean reformulation in terms of the

[43] In fact, this is in basic accord with Derrida's reconceptualization of "the reality of the external world" as radical alterity, on which I commented in the previous section.

[44] As Plotnitsky summarily notes, George Bataille has introduced the idea of "general economy" as a framework "by means of which one can relate to the production, material or intellectual, of excesses that cannot be utilized." In contrast to "restricted economy" based on meaningful production and exchange, general economy is concerned not only with the loss of "excessive

"trace" or "arche-trace."[45] As we saw in the previous section, Derrida's notion of the trace designates the general structure of referral to the other or "radical alterity" as constitutive of what appears to be present (Derrida 1972b, 21/21). In Plotnitsky's account, Bohr's complementarity "proceeds from the empirical – photographic – trace to the general economic, Derridean trace" (1994, 49; cf. 91ff.). For example, when a sequence of water droplets in the cloud chamber serves as a trace of an electron, this trace does not point to something originally present, some independent reality existing behind the trace. Rather:

> One deals here only with traces, traces of traces, and photographs of traces as the effects of a certain *différance*-like efficacity [. . .] (Plotnitsky 1994, 94; cf. 104, 108).

For the "loss of knowledge" due to the uncontrollable object–instrument interaction prevents us from speaking of anything other than such a series of traces or "the complex networks of differential substitutions they entail" (1993a, 8).

Furthermore, continues Plotnitsky, just as Derrida's notion of the trace, the above trace-like structure in Bohr's complementarity also proves to have an aspect of temporalization. Plotnitsky discusses this with specific reference to what is known as the "delayed-choice experiment" (1994, 104; cf. 97), the crux of which was in effect first conceived by Bohr himself. Supposing an electron passing through a slit in a diaphragm, for example, Bohr points out that, even after the passage of the electron, "we are [. . .] still left with a free choice" as to whether we wish to know its momentum or its initial position (PWNB, 4:77). The same holds true as to whether we wish the object to behave as a particle or as a wave. What is essential here is that the behavior of the object is determined afterward depending on the choice thus "postpone[d] [. . .] until a later moment" (PWNB, 2:57), and thus that there occurs an apparent inversion of the normal order of time. This implies, according to Plotnitsky, that, in the delayed-choice experiment, the phenomenon is registered only as a "trace or effect at a certain point, [. . .] the efficacity of which can never – either in the past, present, or future – be seen as *present*" (1994, 105; cf. 108f.). This structure of belatedness is particularly analogous to Derrida's notion of the "supplement" (Plotnitsky 1994, 110), which concerns the circumstance that "by delayed reaction, a possibility produces that to which it is said to be added on" (Derrida 1967c, 99/89). This apparent 'origin' produced as an aftereffect is neither a full presence nor a mere absence. Further, continues Plotnitsky, the above delayed-choice experiment is not an exceptional case, but "only manifests what is found in all quantum experiments,"

energy" on the material level, but also with the loss of meaning "in any interpretive or theoretical process" (1994, 19f.; cf. 1, 29, 34). See Bataille (1967). Plotnitsky characterizes quantum mechanics, and specifically its Bohrian interpretation, as "a general economy of physics," thus situating complementarity within the stream of thought stretching from Nietzsche through Bataille to Derrida (Plotnitsky 1994, 10, 73; cf. 8, 30, 84, 89, 99, 118). He nevertheless focuses on the relation between Bohr's and *Derrida*'s thought, because the latter may be viewed as "the most radically anti-epistemological application of the principles of general economy" (1994, 19; cf. 2).

[45] See Derrida's reading of Bataille's general economy in "De l'écomonie restreinte à l'écomonie générale: Un hegelianisme sans réserve," in 1967a, 369–407/251–77. See also Llewelyn (1986, 10ff.).

namely "the structural, irreducible deferral of presence or absence" of the trajectory. In this sense, one can even say that "[a]ll quantum experiments are delayed-choice experiments," and that "we can only observe delayed traces" (1994, 110, 137). This being the case, the Derridean supplement is relevant to all quantum-mechanical situations, at least in their Bohrian interpretation, and Bohr's complementarity may itself be characterized as "supplementary" (1994, 51; cf. 4; 1993b, 23).

We have so far seen how Plotnitsky's new approach to Bohr's complementarity reveals significant conceptual convergences with Derridean deconstruction. As Plotnitsky puts it, Bohr's thought is not only incompatible with "any form of [...] the metaphysics of presence or ontotheology" (1994, 4), but, in parallel with Derrida's work, "enacts a powerful critique or deconstruction of both classical physics and classical metaphysics" (1994, 1; cf. 5, 66). This suggests the possibility that Derridean deconstruction may also, conversely, be approached from a Bohrian point of view. In fact, Plotnitsky seeks to show, though less extensively than the other way round, how "Derridean configurations may be seen in terms of complementarity" (1994, 209). As we saw in the previous section, in undoing and transforming the metaphysics of presence, which revolves around a series of hierarchical binary oppositions such as self/other, inside/outside, presence/absence, speech/writing, and so on, Derrida introduces the notion of *différance* as the originary movement of differing and deferring, along with accompanying structures such as the trace and supplement mentioned above (see 1994, 38). Specifically, he conceives this *différance* as "the common root of all the oppositional concepts" in such a way that one of the terms in binary opposition "appear[s] as the *différance* of the other, as the other different and deferred" (1972c, 17/9; 1972b, 18/17; cf. Plotnitsky 1994, 40). In light of the commonalities of both thinkers' ideas discussed so far, Plotnitsky characterizes these paired terms as "complementary effects," effects of an operation he calls "complementarization" (1994, 46, 132; cf. 138, 195ff., 232). For, in Derridean deconstruction, those pairs of notions which have traditionally formed hierarchical binary opposites are reconceptualized – similarly to Bohr's case – as "heterogeneously interactive and interactively heterogeneous." That is, Derrida's deconstructive operation on oppositional terms is analogous to Bohr's transformation of such conceptual pairs as subject/object, inside/outside, continuity/discontinuity, and so forth (see 1994, 108, 112, 230).

Although Plotnitsky's analysis as described so far primarily focuses on the affinities and convergences between Bohr's complementarity and Derridean deconstruction, he also examines their "differences" in some respects (1994, 191; cf. 3). In his account, one of the major differences, albeit largely concerning the "balance of emphasis" (1994, 197), may be characterized as the difference between "indeterminacy" and "undecidability." Derridean deconstruction can be seen as an exploration of "undecidability," a term Derrida introduces by analogy to Gödel's incompleteness theorem (Derrida 1962, 39/53; see Plotnitsky 1994, 198). Deconstruction orients itself not primarily toward indeterminacy, but rather toward "the construction of undecidable or aporetic configurations" (1994, 204f.; cf. 112, 207, 229). In fact, Derrida himself, in one of his later works, explicitly "differentiates indeterminacy and undecidability and dissociates [...] his analysis from the framework of indeterminacy" (Plotnitsky 1994, 209; cf. 197): Deconstruction concerns itself with "undecidability," which "is always a determinate oscillation between possibilities," and "[t]hese possibilities are themselves highly determined in strictly defined situations" (Derrida 1990, 273/148). In contrast, continues Plotnitsky, Bohr's complementarity implies "indeterminacy" rather than simply undecidability (1994, 3).[46] To be sure, "in Bohr's matrix, quantum indeterminacy is theoretically determined, no less [...] than mathematical undecidability is determined in Gödel or deconstructive undecidability in Derrida" (1994, 210). Nevertheless, "certain configurations, such as joint measurements or definitions of conjugate variables, can never be fully determined or be seen as determinate or determinable."[47] Further, as an extended theoretical matrix, Bohr's complementarity implies unavoidable indeterminacy "at the level of meaning" in general (1994, 211). In this regard, Plotnitsky assesses Bohr's philosophical orientation as more radical than Derrida's. That is, in Plotnitsky's view, Bohr's complementarity may not simply be conjoined with Derridean deconstruction, but can serve as "a critique of certain, possibly residually metaphysical, aspects of deconstruction" (1994, 5).[48]

There is one more aspect of the Bohr–Derrida relationship that Plotnitsky discusses fairly extensively: a similarity between Bohr's view of classical concepts and Derrida's notion of "closure." As we have seen earlier, it is pivotal to Bohr's complementarity argument that the classical concepts cannot be discarded in quantum theory despite the breakdown of the classical mode of description. Rather, one must essentially be dependent on the use of the classical concepts – not only the concepts specific to classical physics, but more generally "ideas underlying our accounts of every-day experience" (NBCW, 6:399). As Plotnitsky notes, Bohr designates this state of affairs by a generalized concept of the "correspondence principle," which "expresses our endeavours to utilize all the classical concepts by giving them a suitable quantum-theoretical re-interpretation" (PWNB, 1:8).[49] This suggests that the relationship of Bohr's complementarity with the classical tradition "has both continuities and breaks," which "[m]irror[s] the structure of complementarity itself" at a metalevel (1994, 147). This complex relation to the classical tradition is reminiscent of Derrida's attitude toward the Western philosophical tradition. As Plotnitsky puts it,

---

[46] For similar remarks, see Plotnitsky (1993a, 11; 1993b, 70f.; 1994, 196, 212).

[47] Plotnitsky prefers the term 'indeterminacy' to 'uncertainty' to designate Heisenberg's indeterminacy (or uncertainty) relations (see 1994, 7).

[48] Plotnitsky is thus partly critical of Derrida's thought insofar as the latter's "*différance* and accompanying [...] structures and efficacities remain too determinate and are, thus, not radical enough" (1994, 216). He makes, however, some qualifications of this critical assessment: Derrida's analysis in some of his works "appears to lead [...] to the indeterminateness, rather than only undecidability of meaning and context," thus coming closer to Bohr's approach (1994, 219). More precisely, some strata of Derrida's texts as against others do "suggest and enact a complementary interplay of chance and necessity, indeterminacy and determination," or "indeterminacy and undecidability" (1994, 218).

[49] See Plotnitsky (1994, 58, 119, 129, 239).

> Bohr's complementarity is [...] the interplay of restricted — classical — and general-economic — quantum — theories, in part by virtue of Bohr's second great principle — the correspondence principle, which finds its analogy in Derrida's notion of the closure of metaphysics (Plotnitsky 1994, 35; cf. 46, 129).

As we saw in the previous section, Derrida holds that, rather than simply breaking with metaphysics, deconstruction is in a way caught within the "metaphysical closure" (Derrida 1967b, 148/99): One cannot avoid using notions belonging to the metaphysical tradition and thus remains dependent on what is deconstructed. In Plotnitsky's view, by virtue of this common structural feature, neither Bohr's complementarity nor Derridean deconstruction is "a simple or uncritical dismissal of classical theories." Rather, they both enact, in different contexts, a "rigorous suspension" of classical theories, that is, "an analytical exposure of their limitations and a refiguring of classical concepts through a general economy" (Plotnitsky 1994, 11).

Plotnitsky partly differentiates, however, the two thinkers' approaches to the thematic of 'closure.' In his account, although Derrida generally proceeds with the operation of "*différance*-dissemination," he does not apply this operator at the level of closure, and therefore does not speak of "the transformations of closure or of the plurality [...] of closures" (Plotnitsky 1994, 243, 228; cf. 226). In fact, in distinguishing between "two styles or strategies of deconstruction, one more continuous (Heideggerian) and the other more discontinuous (Nietzschean)" in relation to the metaphysical tradition, "Derrida prefers to shift the balance toward the first (more Heideggerian) alternative, while portraying the second as being more likely to reinstate metaphysics" (1994, 232f.). This leads to a certain "globalizing" or "totalizing" tendency of his notion of closure (1994, 247, 259; cf. 261). Plotnitsky argues that one needs "a more complex economy of closure" in which different closures are seen as "complementary in the extended sense of this study" (1994, 245, 240; cf. 265, 267). In other words, reference to Bohr's complementarity can potentially serve to modify Derridean closure into a more pluralist and transformational conception. With this modification, one may speak, for example, of "the closure of discontinuity, along with and complementary to the closure of continuity, or the closure of difference or exteriority complementary to the closure of presence, proximity, similarity, or identity" (1994, 231; cf. 119f.). In this way, "we must radically pluralize or complementarize and allow for transformations of theoretical and, perhaps even, interpretive closures" (1994, 267).

In the present section, surveying Plotnitsky's reading of Bohr's complementarity and Derridean deconstruction, we have seen how his analysis reveals remarkable affinities as well as some divergences between the two thinkers' approaches.[50] His work on this subject may be viewed as a major contribution toward situating Bohr's thought in a broad contemporary philosophical context, extending beyond the frames of the 'mainstream' philosophy of science. It is simultaneously a unique

---

[50] In his more recent writings, Plotnitsky employs the term "singularity," one of the key terms of Derrida's later work, to characterize Bohr's concept of the phenomenon (see Plotnitsky 2002, 70ff.). As noted earlier, however, he no longer directly thematizes the relation between Bohr's and Derrida's thought.

endeavor to broaden the field of discourse on, and associated with, Derridean deconstruction through its interaction with modern physical-scientific thought. This does not mean, however, that all significant aspects of the relation between the two thinkers' ideas have thereby been addressed or clarified. Rather, Plotnitsky's *Complementarity* and related works should be better seen as the beginning of a new mode of inquiry, which will potentially go beyond the thematic already developed in them. In the rest of this chapter, I accordingly wish to examine, through partial critique of Plotnitsky's analysis, a series of as yet unsettled questions concerning the Bohr–Derrida relationship.

6.3 Critical Appraisal of Plotnitsky's Analysis

Notwithstanding its innovative contribution, Plotnitsky's inquiry into Bohr's complementarity and Derridean deconstruction appears to have left not a few unsolved problems. In what follows, I wish to investigate several points of Plotnitsky's account that seem to be unclear or questionable and yet, or rather for that reason, deserve close inspection.

(1) Let us first consider the relation between Bohr's own account of complementarity and Plotnitsky's extension thereof, with regard particularly to the notion of 'mutual exclusion.' We have repeatedly seen that by complementarity Bohr generally means 'mutual exclusion' as well as 'joint completion.' Both these senses are relevant to his overall framework, although, as noted in Section 4.1, the concept of mutual exclusion is used differently in the static and the dynamic conceptions.[51] However, as we saw in the previous section, Plotnitsky modifies Bohr's "definition"[52] of the term in such a way that complementary constituents are not always, but only "at times" mutually exclusive (1994, 75, 73).[53] Moreover, he employs extensively this modified notion of complementarity, not least in his account of the Bohr–Derrida relationship. In particular, differentiating between the Bohrian notion of 'indeterminacy' and Derrida's 'undecidability,' Plotnitsky refers to the latter's characterization of deconstruction as a joint operation of 'either or' and 'neither nor' (see Derrida 1972c, 59/43) and critically points out that "it is this "either or" that marks the space of difference between deconstruction – as undecidability – and complementarity" (1994, 206). This comparison would not hold, however, from

[51] As Catherine Chevalley comments, the notion of mutual exclusion or "incompatibility" is generally constitutive of Bohr's concept of complementarity, except for the complementary relation between different "human cultures," which he does not regard as mutually exclusive. Bohr is, moreover, cautious enough about this exception to call such human cultures only "in a certain sense [. . .] complementary" (PWNB, 2:93). Catherine Chevalley, "Glossaire," in Bohr (1991, 345–567, on 402).

[52] As suggested in Chapter 2, it is the question not of Bohr's *explicit* definition of the term, but rather of an interpretive reconstruction of what he means by the term in various contexts.

[53] In his more recent texts, notably *The Knowable and the Unknowable* (2002) and *Reading Bohr* (2006), Plotnitsky seems largely to refrain from such a modification of the notion of complementarity. See also Plotnitsky (2001b).

the point of view of Bohr's original notion of complementarity, of which 'mutual exclusion' – 'either or' – constitutes an essential part.

Plotnitsky nevertheless appears to hold that although this modification of the concept of complementarity is indeed a step beyond Bohr's explicit remarks, it still remains in accordance with his basic conceptual orientation. To cite a relevant passage, which I earlier quoted only in part:

> Within the scheme just delineated [of a broader complementarity theory], complementary constituents are not always mutually exclusive, as Bohr's definition cited earlier. At a certain level, however, the same is the case in Bohr's complementarity, viewed as a general matrix and overall interpretation of quantum physics (Plotnitsky 1994, 75).

As noted earlier, in speaking of Bohr's "general matrix," Plotnitsky has in mind not only what he calls the "standard" forms of complementarity, specifically the complementarity of space-time coordination and the claim of causality as well as particle-wave complementarity (1994, 121), but further a series of dual (and multiple) notions such as continuity and discontinuity, chance and necessity, interiority and exteriority, and so on. We have also seen, however, that this latter series of notions are designated as complementary by Plotnitsky and not by Bohr. Since these notions, or at least some of them, are conceived in Bohr's work as not always mutually exclusive, the fact that he does not call them complementary is consistent and correlative with his very 'definition' of complementarity. This gives rise to the following questions: What exactly is the reason why Plotnitsky proposes calling those notions (chance and necessity, inside and outside, etc.) complementary in deviation from Bohr's terminology? Does this reason itself still reside in Bohr's own framework or "general matrix" of complementarity? In what way can we regard Plotnitsky's modification of the notion of complementarity not simply as an arbitrary change from without, but rather as an appropriate extension of Bohr's thought, or as a step beyond yet along the lines of his work? It seems to me that Plotnitsky's account itself hardly offers an adequate answer to these questions.

(2) While our subject above revolved around Plotnitsky's stated attempt to extend Bohr's complementarity, I now turn to an implicit yet potentially more significant difference between the two authors' ideas, a difference which will further prove to have direct bearing on the understanding of the Bohr–Derrida relationship. As we have seen earlier, Bohr's idea of complementarity – except for the static-symmetrical conception – is such that the paired relata are not only mutually exclusive, but also contrasting in character, which he expresses metaphorically in terms of the roles of "spectators" and "actors." In other words, at least in certain of its major forms, complementarity revolves around the contrast between the 'spectator's' detachment from phenomena and the 'actor's' involvement in them. Grasping his idea of complementarity thus requires us to recognize which of the two relata – space-time coordination and the claim of causality, for example – corresponds to the 'spectator's' side and which to the 'actor's,' even if this relation of correspondence may change or reverse itself under certain conditions. What is crucial here is that Bohr's approach may generally be characterized as a project of questioning the classical notion of the pure 'spectator,' who would obtain knowledge of the world

without interfering in it, and of indicating how this notion is undermined by the irreducible moment of the 'actor.' That is, he aims at dismantling the privilege hitherto accorded to the 'spectator's' detachment over the 'actor's' involvement, thus operating in a *specific direction* against the hierarchical order of 'spectator/actor.' It is important to note that this operation – to borrow Plotnitsky's term, the operation of "complementarization" – consists in neither simply reversing the hierarchical order nor making the paired relata interchangeable, but reconfiguring their relation in a non-neutralizing manner which I will further specify later.

To be sure, when Plotnitsky primarily stresses Bohr's break with realism while at the same time warning against the anti-realist interpretation, his discussion proceeds practically in line with the above direction in which Bohr's approach operates. For since the 'spectator' posits an object as independent reality, whereas the 'actor' conceives the object as appearing only in its relation to him/herself, Bohr's way of rejecting both realism and anti-realism, as it is discussed by Plotnitsky, may be considered as a critical operation on the hierarchical order of 'spectator/actor.' Plotnitsky does not, however, appear to concern himself systematically with the 'spectator–actor' relation as a notion pivotal to Bohr's complementarity,[54] or with the way in which the above directional character of his operation is constitutive of his very *concept* of complementarity.[55] When, as we have seen earlier, he rightly characterizes the complementarity of space-time coordination and the claim of causality as "dissociating that which is always united" in classical physics, for example, he pays little attention to the contrasting character of the paired relata such that one of the relata serves as an involved 'actor,' and the other as a detached 'spectator.'[56] This largely remains the case when Plotnitsky extends the idea of complementarity to a wide range of conceptual pairs such as continuity and discontinuity, chance and necessity, interiority and exteriority, and so on. Furthermore, if we refer again to one of his most generalizing formulations of complementarity – "diverse [. . .] configurations, double or multiple, operative within the same framework, but without lending themselves to a full synthesis" – we can see that this phrase also makes no mention of the contrast between paired relata or of the specific direction in which their relation is reconfigured so that the traditional hierarchy of 'spectator/actor' is dismantled. Rather, the phrase indicates that Plotnitsky generalizes the idea of complementarity in such a way that these specific characteristics are *not* constitutive of the concept. This also accounts for the apparent ease with

[54] When Plotnitsky at one point draws attention to Bohr's metaphor of 'actors' and 'spectators,' his concern is with its *theatrical* character, rather than with the specific way in which the contrast of 'spectator–actor' is linked with the relation of complementarity. He also makes reservations about this metaphor by saying that "[o]ne must be careful [. . .] to respect the limits of such dramatic or graphic metaphors" (1994, 92).

[55] As may be seen from my discussion in Chapter 3, this applies not only to Plotnitsky, but to most commentators on Bohr's thought.

[56] I argued in Section 4.1 that, in the static-contrastive conception, space-time coordination serves as an involved 'actor,' and the claim of causality as a detached 'spectator,' while, in the dynamic conception, these two complementary relata alternately play the 'actor's' and 'spectator's' roles.

which he extends complementarity from dual to triple or multiple relations. For, with his understanding of complementarity just characterized, it is no wonder that he does not see the necessity of considering the intricate problems of directionality that would inevitably arise with such an extension of complementarity beyond two-term relations. All this suggests that Plotnitsky tends to treat the concept of complementarity not as a relation of 'spectator–actor' contrast, but as a *symmetrical* relation in which the complementary relata appear as virtually interchangeable.

This issue, concerning the interpretation of Bohr's thought, proves to become even more crucial when it comes to intersections with Derridean deconstruction, which essentially deals with the system of paired notions that have traditionally been placed in hierarchical binary opposition. As we saw in Section 6.1, according to Derrida, the operation of deconstruction proceeds through two phases, namely, an overturning of the "order of subordination" and a "general displacement of the system" (1972b, 392/329). He specifically notes that the "phase of overturning" is necessary, because otherwise "one might proceed too quickly to a *neutralization* that *in practice* would leave the previous field untouched" (1972c, 57/41). That is, for deconstruction to be a critical and transformative engagement with the hitherto dominant hierarchy, it cannot be a simple neutralization of the opposing terms. Rather, similarly to, and more systematically than, Bohr's complementarity, it operates in such a specific direction as to undo the hierarchical system.

Plotnitsky, in his lucid account of Derridean deconstruction, indeed generally does justice to its specific directionality as just described. At times, however, as he seeks to connect Bohr's and Derrida's thought, his treatment of Derrida's ideas seems to be affected by *his* reading of Bohr in such a way as to focus less on the directionality of the former's approach as well. Let us look at an instance of this which takes place in connection with a passage from Derrida's *Marges de la philosophie* [*Margins of Philosophy*]. To reproduce a major part of the passage cited by Plotnitsky:

> [. . .] one could reconsider all the pairs of opposites on which philosophy is constructed [. . .] in order [. . .] to see what indicates that each of the terms must appear as the *différance* of the other, as the other different and deferred in the economy of the same (the intelligible as differing-deferring the sensible, as the sensible different and deferred; the concept as different and deferred, differing-deferring intuition, culture as nature different and deferred, differing-deferring; all the others of *physis* — *tekhnē*, *nomos*, *thesis*, society, freedom, history, mind, etc. — as *physis* different and deferred, or as *physis* differing and deferring. *Physis* in *différance*. And in this we may see the site of a reinterpretation of *mimēsis* in its alleged opposition to *physis*) (Derrida 1972b, 18/17).

Derrida's examples here obviously concern the hierarchical binary oppositions of sensible/intelligible, intuition/concept, nature/culture, and so forth.[57] Admittedly, in a way different from oppositions such as presence/absence and present/non-present, the above examples appear to be such that the hierarchical order of each opposition is not generally fixed, but may depend on one's philosophical standpoint – say,

[57] In a parallel passage elsewhere, Derrida speaks of "oppositional concepts that mark our language, such as [. . .] sensible/intelligible, intuition/signification, nature/culture, etc." (1972c, 17/9).

whether empiricist or rationalist – within the metaphysical tradition. Yet, in a *given* philosophical context, the two terms of each opposition can by no means be freely interchanged. What is crucial here is that, as noted in Section 6.1, Derrida expresses the hierarchical binary oppositions (A/B) systematically in the scheme in which the devalued term B appears as the *différance* of the privileged one A (or as A *différé-différant*), and not the other way round.[58]

Immediately following the above citation, however, Plotnitsky comments:

> This economy is reminiscent of Bohr's complementarity, not only insofar as the latter refers to physical representation and variables — such as waves and particles, position and momentum, or coordination and causality — but as it, as a matrix, employs more general concepts and conceptual structures — such as subject and object, interiority and exteriority, analysis and synthesis, concept and intuition, *physis* and *mimēsis*, and so forth — which Bohr sees as complementary as well (Plotnitsky 1994, 40).

Here Plotnitsky hardly seems to focus on the specific hierarchical order in which these pairs of notions have been conceived by metaphysics or on the specific direction in which that order is displaced and disorganized. Let us, in particular, attend to the order in which he writes the conceptual pairs "interiority and exteriority," "concept and intuition," and "*physis* and *mimēsis*," and compare this with Derrida's above passage. In Derrida's account – at least in the above context – it is *intuition* that is metaphysically privileged over the concept, just as inside over outside, or *physis* over its others including *mimēsis*. To be sure, Plotnitsky's passage in question refers not directly to Derrida's views, but to Bohr's matrix of complementarity as associated with Derrida's account quoted earlier. If, however, this connection were drawn with full attention to the above conceptual order in Derrida's account, one would write 'intuition and concept' – rather than conversely – along with 'interiority and exteriority' and '*physis* and *mimēsis*,' or, otherwise, offer the reason for not doing so. This suggests that Plotnitsky's insufficient concern with the directionality of Bohr's 'complementarization' may have occasioned him, at least as regards the above passage, to pay less attention to the directionality of Derridean deconstruction as well. His attempt here to connect complementarity and deconstruction thus appears to result in a certain degree of neutralization and symmetrization of the pairs of notions dealt with by the two thinkers.

---

[58] This might seem to be at odds, however, with the earlier part of the citation: "*each* (my emphasis) of the terms must appear as the *différance* of the other, as the other different and deferred." This clause would suggest that the two terms of each opposition are interchangeable so that, conversely, A can also appear as the *différance* of B. In my view, this apparent inconsistency arises from the English translation (by Alan Bass) which is misguided at a minute yet crucial point. The French original reads: "une nécessité telle que *l'un* des termes y apparaisse comme la différance de l'autre, comme l'autre différé" (my emphasis), which should more appropriately be rendered as: "*one* of the terms must appear as the *différance* of the other ..." That is, the translation of "l'un" by "each" has blurred the specific order of dependence in the metaphysical system and thereby also the specific direction in which that order is deconstructed. This problem of translation might have affected Plotnitsky's reading of the passage, although he refers to the French original as well as the translation.

In discussing Bohr's complementarity in this subsection (2), I have restricted myself to his two conceptions: the static-contrastive and the dynamic conceptions. If, however, we turn to his third, static-*symmetrical* conception, in which the complementary relata appear as similar and interchangeable, we will find out that Plotnitsky's above interpretive tendency is no longer misguided as to this version of complementarity, while on the other hand faced with a problem concerning the relation to Derridean deconstruction. I wish to address this issue in my overall account of the Bohr–Derrida relation in Section 6.4.

(3) Let us now proceed to another set of intersections between Bohr's complementarity and Derridean deconstruction which Plotnitsky points out in terms of a series of Derridean notions such as 'radical alterity,' the 'trace,' the 'supplement,' and '*différance*.' As we have seen earlier, Plotnitsky argues that, in going beyond both realism and anti-realism, Bohr's thought orients itself toward radical alterity and the trace or, more specifically, the structure of the 'delayed trace.' It may be readily noticed, however, that Plotnitsky's account here indeed deals with Bohr's "matrix" of complementarity, but not thematically with his *concept* of complementarity. For this reason, it remains rather unclear how his concept of complementarity entails, or is correlated with, his double break with realism and anti-realism, and in what way it is linked with the above Derridean notions. Further, how is it possible that, as Plotnitsky maintains, a double rejection of realism and anti-realism represents neither a simple contradiction nor an eclectic mixture of the two extremes, but a radical critique of the "metaphysics of presence" in which both are considered as caught up? And how, specifically, is this the case with Bohr's complementarity? Since I presented my own views on part of these questions in Chapter 4, especially in Section 4.2, and will revisit them in Section 6.4, I will here limit myself to examining a few more specific points of Plotnitsky's argument.

First, as noted earlier, in his attempt to connect Bohr's thought with the Derridean notion of the trace, Plotnitsky argues that, in Bohr's interpretation of quantum theory, one deals "only with traces, traces of traces." Here there arises the question: Why and in what way does Bohr's rejection of independent reality lead to such a series of traces, rather than simply to, say, the conceptual construction of an atomic object in an anti-realist manner? While Plotnitsky's account is not explicit on this point, it may be provisionally commented on by recourse to Bohr's remarks in the 'middle' period. As I have repeatedly cited, Bohr wrote in 1949: "any attempt of subdividing the phenomenon will demand a change in the experimental arrangement introducing new possibilities of interaction between objects and measuring instruments which in principle cannot be controlled" (PWNB, 2:40).[59] This may be paraphrased in terms of the trace as follows: Any attempt at determining a trace, with its mode of referral to an 'object,' will introduce a new trace in such a way as to alter the initial referential structure. For this reason, no trace, no mode of

[59] This passage of Bohr is discussed by Plotnitsky (2002, 71), but not in connection with the notion of the trace.

referral, can be determined unambiguously, but every trace is subject to the possibility of a series of traces uncontrollably interfering in each other. This Bohrian notion of the 'trace' indeed has in common with the Derridean trace that any trace may be subject to interactions with other traces, and that no trace refers to anything originally present behind the traces. It seems to me, however, that there are certain limits to this line of analogy. The chain of traces implied by Derrida is such that what is referred to by a trace serves itself as a trace of something else, which in turn serves as a trace of something still else, and so forth. It is not immediately clear, however, whether this applies to Bohr's case. In what sense can one speak of "traces, traces of traces" in such a way that anything referred to by a trace serves itself as a trace of something else? Specifically, if we take again the example in which a sequence of water droplets serves as a trace of an electron, does this electron – or its position or momentum – in turn serve as a trace of something else? Or, instead, does the term 'electron' no more than designate the whole configuration of the traces, in which case it might not be appropriate to speak of traces *of* an electron? These questions appear to suggest the need to delimit more rigorously the structural commonality between the two thinkers' approaches.[60]

Second, as we have seen earlier, when seeking to show the temporalizing character of the Bohrian 'trace' by the example of the delayed-choice experiment, Plotnitsky extends the notion of 'delayed-choice' itself, claiming that "[a]ll quantum experiments are delayed-choice experiments" (1994, 110). In my view, however, he does not fully explicate this generalized notion of 'delayed-choice,' or how it is related to the ordinary sense of the term. In a passage on which Plotnitsky draws, Bohr states that "it obviously can make no difference, as regards observable effects obtainable by a definite experimental arrangement, whether our plans of constructing or handling the instruments are fixed beforehand or whether we prefer to postpone the completion of our planning until a later moment when the particle is already on its way from one instrument to another" (PWNB, 2:57). What he means here is that the way in which the experimental arrangement conditions the phenomenon is not affected by whether or not the arrangement is fixed before the traveling of the particle or, more generally, the proceeding of the 'phenomenon-to-be.'[61] It goes without saying that Bohr does not mean – and it would be simply wrong to suppose – that the choice of the experimental arrangement is always or necessarily delayed in relation to the phenomenon-to-be. This being the case,

---

[60] At one point in *Complementarity*, Plotnitsky expresses reservations about the extent to which the Derridean trace and associated notions are viable: In reference to a passage in *Marges de la philosophie*, he remarks that "[i]f we want 'to go further toward naming' or 'unnaming' what is at stake here, we would have at certain points to suspend this Derridean process [of *différance* and tracing] as well [. . .]" (1994, 256). He seeks to amplify this point in 2001a, esp. p. 68.

[61] Introduced by John Archibald Wheeler in his discussion of the delayed-choice experiment, the term "phenomenon-to-be" refers to that which is not yet a phenomenon, but will become one upon an "act of detection." John Archibald Wheeler, "Law without Law," in Wheeler and Zurek (1983, 182–213, on 183). Commenting favorably on the choice of this term, Plotnitsky connects its implications with the Derridean thematic of the trace (1994, 105).

Plotnitsky's extension of the notion of delayed-choice to all quantum experiments would at least require specifying what 'choice' is delayed in relation to what, and in what way, or else need a qualification such as: 'all quantum experiments are *potentially* delayed-choice experiments.' It appears to me that we could proceed in some other way to make a more suitable comparison between Bohr's complementarity and the Derridean structure of delay or deferral, which I wish to attempt in the next section.

In this section on Plotnitsky's inquiry into complementarity and deconstruction, I have examined a series of philosophical problems apparently involved in or left unsolved by his analysis. It seems to me – and has in part already been suggested – that underlying some of these problems is his insufficient concern with both the historical and the structural complexity of Bohr's complementarity. While discussing different (standard as well as non-standard) forms of complementarity, Plotnitsky does not consider heterogeneous *conceptions* of complementarity – such as those I designate as static and dynamic – that would imply different and perhaps incompatible philosophical orientations. This also appears to be linked with his inadequate attention to diachronic changes in Bohr's thought, which, in my view, may be analyzed in terms of such different conceptions of complementarity.[62] In the next and final section, drawing on my earlier analysis (in Chapter 4) of these philosophical and historical problems regarding Bohr's work, I wish to outline my basic approach to the relation between Bohr's complementarity and Derridean deconstruction. In so doing, I will also return to some issues concerning Plotnitsky's account that have not been fully examined in this section.

6.4 Complementarity and Derridean Deconstruction

Our inquiry into the conceptual intersections between Bohr's complementarity and Derridean deconstruction will, in a sense, be less straightforward than our preceding account of the relation to hermeneutic philosophy. For, unlike hermeneutics, Derridean deconstruction is not centrally concerned with a conceptual pair (such as belonging and distanciation) that would be directly parallel to Bohr's notion of complementarity as the relation between the roles of 'actors' and 'spectators.' This does not mean, however, that his thought has less in common with Derridean deconstruction than with Gadamer's or Ricoeur's hermeneutics. As we have repeatedly seen, Bohr's idea of complementarity may generally be characterized as a project of questioning and dislodging the classical notion of the pure 'spectator' by indicating the irreducible moment of the 'actor.' This critique of the privilege hitherto accorded to

---

[62] Admittedly, like some other commentators, Plotnitsky has recently come to recognize more clearly a certain historical change in Bohr's thought. While in *Complementarity* largely treating his post-EPR development as a 'refinement' of the mode of presentation, Plotnitsky now characterizes the same process as a substantive advancement toward a fully "nonclassical" formulation of complementarity (2002, 36, 58ff.; cf. 2003, 1660, 1663; 2006, 17ff.). Nevertheless his views on the diachronic dimension of Bohr's thought still seem to miss some of the points I hold to be essential.

the 'spectator's' detachment over the 'actor's' involvement appears to constitute the most basic point of contact with Derridean deconstruction: It may indeed be viewed as a Bohrian intervention in the hierarchical binary opposition of 'spectator/actor' as part of the metaphysical system. This linkage becomes more complex, however, as we distinguish different conceptions of complementarity in Bohr's thought. In what follows, based on my analysis in Chapter 4 of these distinct – static-contrastive, dynamic, and static-symmetrical – conceptions, I will accordingly examine whether and, if so, how each of them may be associated with Derridean deconstruction.

(1) As we have seen, Bohr's *static-contrastive* conception of complementarity, which is represented by his 'early' idea of the relation between space-time coordination and the claim of causality, simply juxtaposes two complementary relata, the 'spectator's' detachment from the phenomenon and the 'actor's' involvement therein. By showing the 'actor's' involvement to be unavoidable and irreducible, this conception of complementarity puts the two opposing moments, detachment and involvement, on an equal footing. Bohr's operation here is thus directed against the hierarchy of the binary opposition, but *not* against the binary framework itself. That is, it still remains within the binary framework of 'spectator/actor,' while depriving the 'spectator' of its traditional privilege. This being the case, the static-contrastive conception may only in a limited sense be compared with part of Derridean deconstruction. As noted earlier, in Derrida's account, the first phase of the deconstructive operation consists in overturning the hierarchical order without immediately going beyond the binary framework. We can see that Bohr's static-contrastive version of complementarity is in a way similar to this first phase of deconstruction, but that it nevertheless differs from the latter insofar as it is restricted to putting the binary opposites on a par instead of reversing their order. Bohr's thought here can be still less likened to Derrida's *overall* idea of deconstruction. While this static-contrastive conception represents a major innovation in the history of scientific thought, its intersection with Derridean deconstruction is thus considerably limited.

(2) In contrast, Bohr's *dynamic* conception of complementarity, which was extended from non-physical fields to quantum theory in the 'middle' period, constitutes the very movement of displacing the system of binary opposition, thus bearing a crucial similarity to Derridean deconstruction. As we have seen, in this conception, the 'spectator's' detached reflection on the 'actor's' involvement proves to have itself an aspect of involvement, thus leading to an indefinitely extensible series of reflections. Here the two complementary relata – the roles of the 'spectator' and the 'actor' – while still conflicting in character, are no longer in binary opposition, but are reinscribed in the series of reflections in which what is reflected upon is each time uncontrollably displaced in meaning. In other words, Bohr's dynamic conception not only disrupts the hierarchical order of the binary opposites, but undoes the system of binary opposition itself – not by neutralizing, but by continually displacing the opposing items so that the reflective determination of phenomena at any point assumes an ambiguous and 'floating' character. This notion of complementarity as dynamic displacement is indeed philosophically so radical that the very designation 'complementarity' might no longer be quite appropriate. It is thus

specifically with regard to this dynamic conception – not to Bohr's whole framework of complementarity – that we can speak of significant conceptual links with Derridean deconstruction.

This linkage should be further specified with respect to the relevance of particular components of Derrida's thought as well. It appears to me that Bohr's dynamic conception of complementarity may be most directly associated with Derrida's critique of the "transcendental signified" – the logocentric notion of the meaning that "in and of itself, in its essence, would refer to no signifier, would exceed the chain of signs" (1972c, 30/19f.). As we saw in Section 6.1, Derrida argues that this notion of the transcendental signified breaks down because every signified itself "signifies again," that is to say, "is also in the position of a signifier," thus leading to an indefinitely extensible series of signs (1967a, 42/25; 1972c, 30/20). Here we can see a specific point at which Bohr's and Derrida's philosophical approaches tend to converge. For if the phenomenon in quantum theory as well as in other fields may be conceived as a 'sign' in a broadened sense, we can paraphrase Bohr's dynamic conception as follows: While the 'spectator' aims at determining the signified meaning of an observed sign, this very attempt serves as an 'actor' in uncontrollably altering the signifying structure. Although this is not strictly the same as Derrida's idea mentioned above, according to which it is the signified itself (rather than an attempt at its determination) that signifies again, the crucial line of thought common to both thinkers may be rendered as follows: Any access to the signified unavoidably carries with it the emergence of a new signifier, and is therefore subject to an indefinitely extensible series of signs in which the signifying structure is each time uncontrollably displaced. In this way, the dynamic version of Bohr's complementarity revolves around an idea closely parallel to a key aspect of Derridean deconstruction, the critique of the transcendental signified.

Starting from this specific point of convergence, we can see how Bohr's dynamic conception may be further associated with other pivotal concepts and ideas of Derridean deconstruction. To begin with, Bohr's above line of thought as shared by Derrida may be linked (at least in part) to the two basic meanings of the latter's term "*différance*": deferring and differing. First, since any attempt by the 'spectator' to determine the signified meaning assumes the 'actor's' role of intervening in the signifying structure, access to the signified as such is indefinitely deferred. Second, by the same token, any attempt at the reflective identification of the signified meaning involves an uncontrollable alteration and differentiation thereof.[63] These two aspects taken together, we can also see how the notion of the "presence" of meaning breaks down: For Bohr as well as for Derrida, the signified meaning can never present itself as such, never be attained in its full presence, because of the above double movement of deferral and differing. An insight into this unavoidable non-presence of meaning appears to underlie Bohr's emphasis on an essential "ambiguity in our use of language," the ambiguity of "even words like 'to be' and 'to

[63] Admittedly, these two points do not cover all the meanings of *différance* conceived by Derrida, which also include, in particular, the production of the Saussurean differential system of signs.

know'" (PWNB, 1:19). Further, the same dynamic line of thought may also be connected with the Derridean concept of "radical alterity." As noted earlier, although Plotnitsky focuses on radical alterity – associated with the double break with realism and anti-realism – in his analysis of the Bohr–Derrida relation, he leaves rather unclear the specific manner in which this notion is implied by Bohr's concept of complementarity. In Chapter 4, I argued that Bohr (except for his static-symmetrical conception) subscribes to neither realism nor anti-realism, but reconceives the very relation between them in the framework of complementarity. Specifically in his dynamic conception, this framework takes the form of an indefinitely repeatable alternation of two conflicting moves, the realist separation of subject/object and the non-realist undoing thereof. What is thus both posited and disrupted in this undecidable alternation may no longer appropriately be called 'objective reality,' but rather be characterized as 'radical alterity' in the Derridean sense. In other words, Bohr's dynamic conception implies the notion of the object as radically other, the other that indeed appears in relation to the self and yet irreducibly exceeds this very relation.

This series of links between Bohr's and Derrida's philosophical approaches do not cover, however, all the major aspects of Derridean deconstruction. In particular, we cannot – or at least cannot immediately – connect Bohr's complementarity with Derrida's thematic of "iterability," one of the basic strands of his deconstructive project. As we have seen, according to Derrida, any linguistic sign can serve as such insofar as it "may be indefinitely repeated as the same" (1967c, 8/9), and, in this sense, repeatability, or rather "iterability," is constitutive of language in general and, by extension, of all experience. This circumstance – the dependence of the identity of a sign or mark on its indefinite iterability – also offers the reason why access to the meaning is indefinitely deferred. As noted earlier, no less crucial to this idea of iterability is that iteration does not simply reproduce the same, but "always alters [. . .] that which it seems to reproduce" (1990, 82/40). This Derridean notion of iterability does not, however, have any direct counterpart in Bohr's complementarity argument.[64] When at times using the term 'repetition' (see PWNB, 3:4, 4:182),[65] Bohr seems to conceive it simply in its ordinary sense – not in a Derridean way as constitutive of the identity of what is repeated, or as implying not only identity, but also difference and alteration. We can thus see that Bohr's dynamic conception indeed significantly converges with a series of aspects of Derrida's thought, but not with his thematic of iterability. This seems also to account for the difference, suggested in the previous section, between Derrida's concept of the trace and the trace-like structure in Bohr's thought. Derrida's notion of the chain of traces – in which what is referred to by a trace serves itself as a trace of something else, which in turn serves as a trace of something still else, and so on – is implied by his idea of iterability insofar as this idea may be conceived as a kind of referral to the earlier

---

[64] If, however, Derridean iterability is reconceptualized with a focus on the iteration of the signified rather than of the signifying form, we might be able to interpret Bohr's indefinitely extensible series of reflections as a kind of quasi-Derridean iteration.

[65] These references are to passages from the 'late' period, which have little to do with the dynamic conception of complementarity.

(or later) occurrences of signs. On the other hand, although some of Bohr's ideas can be rendered in terms of the trace in the sense of referral to the other, his thought does not have a notion equivalent to the Derridean structure of traces, because of its lack of an essential link with the idea of iterability.

Yet, notwithstanding the above delimitation, the conceptual linkage between Bohr's complementarity and Derridean deconstruction may in a certain sense be viewed as closer than the relationship with hermeneutic philosophy as discussed earlier. To begin with, as I pointed out in Section 5.3, Bohr's dynamic conception differs (despite significant similarities) from Ricoeur's concept of dialectics, whose polar moments are each identified as such and put in order as definite stages to form a "hermeneutical arc." Unlike this hermeneutic idea, Bohr's and Derrida's orientations both imply a series of signs in which signified and signifier – or the roles of 'spectators' and 'actors' – are each deprived of fixed self-identity and are subject to potentially interminable displacement and alternation. Further, and linked with the above, I have also noted that Bohr does not privilege the 'actor's' involvement over the 'spectator's' detachment, whereas Gadamer and Ricoeur give priority to belonging over alienation or distanciation. In this respect, too, Bohr's dynamic conception has more in common with Derrida's approach: While seeking to dislodge the privilege traditionally accorded to one of two opposing concepts (in Bohr's case, the 'spectator's' detachment), both Bohr and Derrida avoid simply reversing the order of dependence and instead proceed to displace and undo the system of binary oppositions itself. These points suggest that Bohr's complementarity, specifically in its dynamic version, may be characterized as a deconstructive – rather than hermeneutic – critique of the hierarchical binarism of metaphysics, indeed of the physical-scientific subsystem of "the metaphysics of presence."

So far, in subsections (1) and (2), I have reexamined Bohr's major conceptions of complementarity in the 'early' and 'middle' periods, and emphasized, in particular, remarkable similarities between his dynamic conception and Derridean deconstruction. Despite this convergence, however, there appears to be one more important difference, or rather a contrast, between Bohr's and Derrida's philosophical approaches, even as regards the former's dynamic conception. While, since the previous section, I have stressed the specific *directionality* – inadequately attended to by Plotnitsky – in which both Bohr and Derrida critically intervene in the traditional hierarchies, their contrast to be discussed here pertains precisely to this directionality. I wish to explicate this in terms of the paired concepts of 'proximity' and 'distance.'[66] As we have repeatedly seen, the primary target of Bohr's epistemological critique is the 'spectator's' detachment, oriented to the separation of subject/object, and may thus be characterized as a moment of *distance*. On the other hand, its complement, the 'actor's' involvement, implies the undoing of this separation and therefore constitutes a moment of *proximity*. In Bohr's thought, it is this

[66] This pair of concepts is extensively used by Plotnitsky in different contexts. Specifically, he characterizes the relation of Derridean deconstruction to Hegelian philosophy in terms of the "complementarity" of "proximity" and "distance" (Plotnitsky 1993b, 90; cf. 1994, 232ff.).

latter moment of proximity that plays the role of displacing and altering the signifying structure and thereby serves to disrupt the hierarchical order of 'spectator/actor.' In contrast, Derrida's critique is directed at the metaphysical tradition primarily characterized by the notion of immediate presence as "absolute proximity" or the reduction of distance. The deconstruction of this metaphysics accordingly proceeds by virtue of 'distancing' moments such as *différance* and arche-writing, oriented toward dismantling the hierarchical binary opposition of presence/absence. In this way, the overall strategic roles of distance and proximity in Bohr's and Derrida's projects prove to be inverse to each other. This is no doubt closely connected with the difference between the respective intellectual contexts in which the two thinkers' ideas were historically formed: Bohr's complementarity was designed primarily to challenge the classical objectivist notion of physical reality, prevalent among physical scientists, whereas Derrida developed his thought in critical engagement above all with Husserlian phenomenology, which rests on a form of transcendental subjectivity. We can thus see how Bohr's complementarity, for all its affinities with Derridean deconstruction, differs from, and forms a contrast with, the latter in the overall directionality of operation.[67] To be sure, both Bohr's and Derrida's thought not only operate in opposite directions through the medium of the opposition of proximity/distance, but, precisely by doing so, tend to undo this very opposition as associated with other hierarchical binary oppositions. This does not mean, however, that the two thinkers' philosophical approaches eventually fully converge with regard to the paired notions of proximity and distance. Rather, their contrast as indicated above may serve to question and potentially go beyond what appears to be their respective one-sided attitudes toward the philosophical tradition or traditions, without thereby readily synthesizing them into a global or comprehensive standpoint.

(3) Let us finally turn to Bohr's third, *static-symmetrical* conception of complementarity, developed during the 'middle' and 'late' periods. Despite their significant mutual differences, Bohr's static-contrastive and dynamic conceptions of complementarity have in common the circumstance that the two complementary relata contrast with each other as moments of being a 'spectator' and an 'actor,' even

[67] As we saw in Section 6.2, Plotnitsky's analysis of the Bohr–Derrida relation contains a partial critique of Derrida's notion of closure. According to Plotnitsky, one can and should speak not only, as Derrida does, of "the closure of continuity, [...] presence, *proximity*, similarity, or identity" (my emphasis), but also complementarily of the closure of "discontinuity, [...] difference or exteriority" (1994, 231). Our consideration here suggests that, to these last series of terms, we could add 'the closure of distance,' with which Bohr's complementarity critically concerns itself. As we have seen, Plotnitsky's above remark is intended to question the unitary and totalizing character of Derrida's notion of closure and to pluralize this notion from a Bohrian point of view. To be sure, insofar as this move of Plotnitsky rests on his general understanding of Bohr's complementarity as operating equally in the two opposite (or more) directions, it does not accord with my point of view. Yet, in the present context, we could reconfigure Plotnitsky's argument in such a way as to characterize the above contrast between Bohr's and Derrida's orientations in terms of the contrast between the different closures – the closures of distance and of proximity – which are correlative with the contrasting directionalities of their respective critical operations.

though this relationship is dynamically displaced. By contrast, as we have seen earlier, in his static-symmetrical conception, each of the complementary relata has the character of being partly a 'spectator' and partly an 'actor,' thus appearing as similar and interchangeable items. This version of complementarity is thus devoid not only of the movement of conceptual displacement, but of any specific directionality of operation, such as the kind of directionality essential to the earlier versions of complementarity. It may rather be characterized as a static and neutralized *product* of the directional operation by which the hierarchical order of 'spectator/actor' is displaced and disorganized – a product that itself no longer embodies this operation. We can thus see that this static-symmetrical conception has little in common with Derridean deconstruction, which, as repeatedly noted, is marked by the specific directionality of its critique of metaphysics. Notwithstanding its theoretical significance as an interpretation of quantum theory, we can hardly speak of the philosophical radicality of this conception or its connections with Derridean deconstruction.

Furthermore, as we have seen, it is based on this static-symmetrical conception that Bohr developed his 'late'-period objectivism, which proceeds with an attempt to restore the standpoint of a pure 'spectator' or, in his own term, a "detached observer" (PWNB, 2:74). By means of the mechanism of 'conceptual containment,' he thus seeks to privilege the standpoint of the 'spectator' over that of the 'actor,' or, in other words, to reestablish the hierarchical binary oppositions of detachment/involvement, objective/subjective, and so forth. From a deconstructive point of view – such as Bohr's own earlier one – this may be characterized as nothing other than a return to the metaphysical tradition. This return is, however, as suggested earlier, not a simple conservative regression, but rather is mediated by the paradoxical process in which his radicalized engagement with the moments of being 'actors' was reversed to an apparent overcoming of these moments.

In my appraisal of Plotnitsky's reading of Bohr in the previous section, I critically pointed out that he does not attend adequately to the specific directionality of 'complementarization,' but tends to neutralize and symmetrize the complementary relationship. This critique indeed holds good as far as Bohr's static-contrastive and dynamic conceptions are concerned. His later, static-symmetrical conception, however, does accord basically with the neutralizing tendency of Plotnitsky's interpretation. In other words, Plotnitsky's reading applies to this final version of complementarity with regard to the above question of directionality or the lack thereof. This being the case, a major problem facing Plotnitsky's analysis appears to be as follows: While Plotnitsky emphasizes the philosophical radicality of Bohr's thought and its close affinity with Derridean deconstruction, a significant aspect of his interpretation is in accordance with the static-symmetrical conception of complementarity, which can in no way be regarded as deconstructive in the Derridean sense.

In this section, with critical reference to the interpretive problems of Plotnitsky's analysis, I have offered an alternative approach of my own to the conceptual intersections between Bohr's complementarity and Derridean deconstruction. We can now see how the linkage between the two thinkers' ideas have thereby been both amplified and delimited – with regard to the overall character as well as the specific aspects and versions of their respective projects. As noted in the beginning of this chapter, in addressing Derrida's thought, I have largely restricted myself to his early work up until the mid-1970s. The relation between Bohr's and Derrida's philosophical orientations might become still more complex if we take into account the latter's reelaboration of deconstruction in later years. In particular, Derrida's reformulation of "iterability" as a paradoxical relation between "the rule and the event" or between "concept and singularity" seems to suggest new elements of intersection with Bohr's complementarity (1990, 215f./119). That is, if complementarity as the 'actor–spectator' relationship, specifically in its dynamic conception, implies a conflictual relation between the determination of meaning and the singular event surpassing the determined meaning, it may have certain affinities with what the later Derrida characerizes as the "aporetic" relations between law and justice, exchange and the gift, universal and singular responsibility, and so forth.[68] This issue, along with potentially many others, must, however, be left for inquiry on another occasion.

To conclude this final chapter, I wish to comment briefly on what is known as the "Science Wars," insofar as it is concerned with our present thematic of complementarity and deconstruction. Around the middle of the 1990s, mainly in the United States but to some extent internationally, there arose a heated debate called the Science Wars between natural scientists (and some philosophers of science), on the one hand, and researchers in science studies and other cultural and philosophical fields, on the other – the former defending the integrity of science perceived as attacked and discredited by the latter. In a sense, this debate has little to do with the content of this chapter or even the whole present book, insofar as one of the major aspects of the debate is concerned: the issue of the alleged misuse or abuse of scientific terms by postmodernist and poststructuralist authors. In fact, the physicists Sokal and Bricmont's 1997 polemical book *Impostures Intellectuelles* [*Intellectual Impostures*] largely spares Derridean deconstruction from their criticism of this kind, noting that, unlike the cases of other French poststructuralists, "there is no systematic misuse of [...] science in Derrida's work" (1997, 43/7).[69]

---

[68] Among Derrida's many other works, see, for example, *Force de loi* (1994), which thematizes the aporetic relation between law and justice. For an exposition of Derrida's thought throughout his career, with special focus on the relation between his earlier and later work, see Takahashi (1998). See also Katsumori (2010), which discusses Derrida's earlier and later thought in connection with the question of nature.

[69] In the earlier phase of the debate, Gross and Levitt as well as Alan Sokal did denounce Derrida along with others for abuses of scientific terms (Gross and Levitt 1994, 75–9; Sokal 1996a, 209; cf. Sokal and Bricmont 1997, 373/244). After and perhaps owing to a counterargument by Roger Hart in his article "The Flight from Reason" (Ross 1996, 259–92, on 277f.), however, they apparently ceased targeting Derrida's work. Derrida himself briefly mentions this course of events in his comments on the debate (2001, 279–81/70–72). Incidentally, in the Science Wars, although the term 'hermeneutics' was at times mentioned by the natural scientists as part of the problematic cultural critique of science and even included in the subtitle of Sokal's hoax paper (see Gross and

The scope of the Science Wars was not restricted, however, to the question of the correct or incorrect usage of scientific terms. Rather, this debate has a more substantive theoretical dimension that does seem to be relevant to the thematic of the present study.

Let us recall that, in his controversial 1996 article, "Transgressing the Boundaries," Alan Sokal cites some passages from Bohr's and Heisenberg's work, including the former's remarks that "[a]n independent reality in the ordinary physical sense can neither be ascribed to the phenomena nor to the agencies of observation," and that "the conscious analysis of any concept stands in a relation of exclusion to its immediate application" (Sokal 1996a, 202f.).[70] He goes on to say that these passages suggest "profound connections between complementarity and deconstruction," which "have recently been elucidated" by some researchers, especially, and "in great depth, by Plotnitsky" (1996a, 204). After revealing this article to be a "hoax," Sokal, jointly with Bricmont, comments on the above account of the linkage between complementarity and deconstruction that it was meant as a parody of postmodernists' "fondness for the most subjectivist writings of Heisenberg and Bohr, interpreted in a radical way that goes far beyond their own views (which are in turn vigorously disputed by many physicists and philosophers of science)" (Sokal and Bricmont 1997, 371/242). In my view, however, apart from the problem of confounding deconstruction and postmodernism or treating the former as just a variant of the latter, this remark indicates a double misconception of complementarity and deconstruction on the authors' part. As we have seen, Bohr's complementarity indeed proceeds with a critique of the notion of objective reality, and yet – in its various formulations including those cited by Sokal – can by no means be characterized as subjectivist. Far from reducing everything to the subjective, Bohr seeks to conceive the relation between the subject/object separation and the undoing thereof in terms of the very idea of complementarity. It should also be obvious from the discussion in this chapter that neither Derridean deconstruction nor Plotnitsky's (or my) analysis of deconstruction and complementarity may reasonably be called subjectivist. The Derridean conception of 'radical alterity,' in particular, which Plotnitsky employs as a key idea in his account of complementarity, is in no way subjectivist, but, on the contrary, focuses precisely on the other or otherness that exceeds the scope of the subjective self.[71]

Levitt 1994, ix, 51, 117; Sokal 1996a, 199), hermeneutic philosophy as such, including the work of Gadamer and Ricoeur, was never made a principal target of their criticism.

[70] Sokal's article "Transgressing the Boundaries: Toward a Transformative Hermeneutics of Quantum Gravity" was originally published in *Social Text*, 46/47 (1996): 217–52, and later reprinted in *Intellectual Impostures* (the English version of Sokal and Bricmont 1997), pp. 199–240. The page numbers indicated here are those in the latter publication. A French translation of the paper is given in Sokal and Bricmont (1997, 305–67).

[71] See Plotnitsky (1997), in which he responds to the Science Wars, seeking specifically to refute Sokal's and Gross and Levitt's criticisms of Derrida.

The label of 'subjectivism' here hardly makes sense unless one overly extends this term to cover a vast range of different philosophical positions, perhaps all positions other than objectivism as embraced by Sokal and other scientists. This is indeed what these scientists in effect do in the Science Wars, and they also similarly use such terms as 'relativism' and 'cultural (or social) constructivism,' which, in their usage, overlap to a large extent with the term 'subjectivism' (e.g. Gross and Levitt 1994, 42; Sokal and Bricmont 1997, 34/2).[72] For instance, after describing himself as a scientist "who believes [. . .] that there exist an external world, that there exist objective truths about that world, and that my job is to discover some of them," Sokal continues that "[i]f science were merely a negotiation of social conventions about what is agreed to be 'true,' why would I bother devoting a large fraction of my all-too-short life to it?" (1996b, 249; cf. Sokal and Bricmont 1997, 92/51, 377/246). This sounds as though there were only two possible philosophical orientations: belief in the existence of objective reality and truth, on the one hand, and the reduction of everything to social constructs, on the other. As we can readily see, Bohr's as well as Derrida's thought essentially evades such a simple alternative.

Linked with the above alternative, Sokal and Bricmont at one point introduce yet another alternative regarding the critique of science: "It is important to distinguish carefully between two different types of critiques of the sciences," namely, "those that are opposed to a particular theory and are based on specific arguments, and those that repeat [. . .] the traditional arguments of radical skepticism." The former type of critique, continue the authors, "can be interesting but can also be refuted," whereas the latter, which questions "the viability of logic or the possibility of knowing the world through observation and/or experiment," is "irrefutable but uninteresting" (1997, 279/179). Uninteresting to whom? To the authors themselves and perhaps to many contemporary scientists, but undoubtedly *not* to Bohr. As my discussion throughout this study suggests, Bohr's complementarity cannot simply be classified into either of the above two types of critiques of science. It indeed proceeds with specific arguments on classical and quantum physics, and yet, precisely through these arguments, develops a radical critique – rather than a radical skepticism – of the conventional views of nature and of our knowledge and experience thereof. This being the case, if Bohr lived today or had lived recently to witness the Science Wars, we could hardly imagine him simply taking Sokal's side. On the contrary, Bohr's complementarity, especially its dynamic conception, calls into question precisely what Sokal and other scientists in the debate take for granted, notably the unambiguity and full conceptual containability of our scientific knowledge. Here we may be struck by an enormous distance in the basic mode of thought between Bohr, on the one hand, and Sokal and other like-minded contemporary scientists, on the other (see Kanamori 2000, 146f.).[73] This is not simply due to Bohr's personal

[72] Here I do not enter into the relatively minor conceptual differences among the natural scientists named here, or more generally among the scientists involved in the Science Wars.

[73] Bohr's 'late'-period objectivism, however, based on his static-symmetrical conception of complementarity, may be considered an initial step toward the forms of objectivism embraced by the scientists involved in the Science Wars.

qualities, but also, and perhaps more essentially, to the historical context in which he lived and developed his thought, and to the subsequent course of events in which quantum physics has fully established itself as a 'normal science' and thereby the philosophical questions involved in the field have become increasingly invisible to the eyes internal to the discipline.[74] A thematic analysis of this issue, however, lies beyond the scope of the present study.

---

[74] Drawing on Thomas S. Kuhn's point of view, von Weizsäcker notes that "normal science" makes "regular progress [...] precisely by not posing perplexingly unresolved basic questions" (von Weizsäcker 1983, 159). See also Barad (2007, 68f., 252f., 275).

Concluding Remarks

In the present study, I have explored Niels Bohr's idea of complementarity with a focus on its structural complexity, historical changes, and intersections with recent and contemporary philosophical approaches, specifically hermeneutics and deconstruction. Following a brief review of the rise of quantum theory in the first chapter, I surveyed in Chapter 2 the development of Bohr's complementarity during his career, and in Chapter 3 a number of prior interpretations of his thought. My own interpretive approach was explicitly developed from Chapter 4 onward. Throughout this analysis, I have attached special importance to Bohr's characterization of complementarity in terms of the relation between the roles of 'spectators' and 'actors,' or, to put it less metaphorically, between detachment and involvement. The outcome of the inquiry may be summarized under the following rubrics: (1) the conceptual structure and (2) the historical development of Bohr's complementarity, (3) some of its further philosophical implications, and (4) the relation of complementarity to hermeneutic philosophy and (5) to Derridean deconstruction.

(1) My analysis, particularly in Chapter 4, has shown that, despite its possible appearance to the contrary, Bohr's complementarity is by no means structurally simple or homogeneous. Abstracting for the moment from the diachronic dimension, we can distinguish at least three different and potentially conflicting conceptions of complementarity: the static-contrastive, the dynamic, and the static-symmetrical conceptions. To begin with, the *static-contrastive* conception refers to the circumstance that one behaves as an 'actor' and as a 'spectator' only in different and separate situations. In other words, complementarity in this sense may be characterized as a onefold contrasting juxtaposition of the 'actor's' involvement in the phenomenon and the 'spectator's' detachment from it. Next, the *dynamic* conception revolves around the relation between the 'actor's' involvement in the phenomenon and the 'spectator's' detached reflection *on that involvement*. While contrasting with involvement, reflection here proves not to leave the phenomenon purely intact, but to introduce an unavoidable interaction with it, that is, to have itself an aspect of involvement, thus leading to an indefinitely extensible series of reflections. This second conception of complementarity may therefore be characterized as a dynamic and multifold displacement of the 'spectator–actor' relationship. Finally, the *static-symmetrical* conception, while juxtaposing two complementary relata, differs from

M. Katsumori, *Niels Bohr's Complementarity*, Boston Studies in the Philosophy of Science 286, DOI 10.1007/978-94-007-1748-0,

the static-contrastive one in that it imparts to each of the relata both moments of being a 'spectator' and an 'actor.' In other words, the third conception treats the complementary relata not as contrasting, but as similar and interchangeable items.

(2) Arguing that Bohr's complementarity underwent significant historical changes far beyond a mere terminological refinement, I have sought to spell out these changes in terms of the above three different conceptions. What I have called the 'early' period, stretching from 1927 to the mid-thirties, is characterized by a coexistence of the static-contrastive and dynamic conceptions. In this period, Bohr's static-contrastive conception of complementarity was exemplified by the quantum-theoretical relation between space-time coordination and the claim of causality – or, more or less equivalently, between observation and state definition – corresponding to the roles of an 'actor' and a 'spectator,' respectively. On the other hand, his dynamic conception was largely restricted to fields outside quantum physics, taking such forms as the complementarity of psychical experience and its reflective analysis, although this conception also played a limited role in his account of quantum theory.

In the 'middle' period, however, starting with his 1935 debate with EPR, Bohr revised his idea of complementarity in quantum theory in such a way that any measurement – with regard to either space-time coordination or momentum-energy conservation – while serving as the 'spectator's' detached reflection on the 'actor's' involvement, turns out to have itself a character of involvement. His transition from the 'early' to the 'middle' period is thus marked by a full extension of the dynamic conception of complementarity to the very field of quantum theory. This expansion of his dynamic mode of thought, however, served paradoxically to produce a new static conception. That is, the double character of each of the complementary relata as involvement and detachment, once abstracted from the dynamic dimension, led to a static-symmetrical juxtaposition of the two relata.

Further, from around 1950 onward, Bohr's dynamic conception of complementarity gave way entirely to his static-symmetrical conception, on the basis of which he developed a 'late'-period objectivist philosophy. He came to assert the thoroughly unambiguous and objective character of the description of nature as offered by quantum theory, stressing, in particular, "our position as detached observers of nature." His reasoning underlying this claim is that complementary description, by reflectively incorporating in itself all moments of being an 'actor,' embodies the standpoint of a pure 'spectator,' that is, a subject that objectifies the two complementary relata without giving rise to any further series of reflections. In this way, Bohr went through a complex philosophical path over the three periods, finally arriving at an objectivist standpoint quite distant from his earlier positions.

(3) I have investigated further philosophical implications of Bohr's complementarity with critical reference to a number of prior interpretations as reviewed in Chapter 3. First, while my above structural analysis of complementarity is partly similar to von Weizsäcker's account of "parallel" and "circular" complementarity, it differs from the latter, which attends to neither the contrast nor the dynamic displacement of the roles of 'spectators' and 'actors.' Second, while many other

commentators have attempted to identify Bohr's philosophical thought with a specific position along the axis of realism and anti-realism, my inquiry has indicated the essential limitations of such interpretive approaches: In the static-contrastive as well as the dynamic conception, Bohr suspends the validity of both realism and anti-realism, and *situates the very relation between them* – or, more precisely, between the realist separation of subject/object and the non-realist undoing thereof – *within the framework of complementarity*. Third, among various studies on the relation between Bohr's and Kant's thought, I have specifically focused on Kaiser's attempt to connect the two thinkers' ideas in terms of the mechanism of "conceptual containment." Although Kaiser's argument points to an important aspect of Bohr's thought, it does not accord with his dynamic conception of complementarity, which, unlike Kant's philosophy, implies that concepts and words – while subject to a movement of containment – *can never be fully contained*. On the other hand, Bohr's objectivist thought in the 'late' period, based on his static-symmetrical conception, implies the necessity as well as the attainability of full containment, thus coming close to Kaiser's point of view.

As indicated in the above divisions (1)–(3), my overall analysis of Bohr's complementarity has emphasized the significance of his dynamic conception, which, initially largely restricted to non-physical fields, came to play a crucial role in his renewed interpretation of quantum theory in the 'middle' period, and yet subsequently receded and disappeared in his 'late'-period objectivism. As it has turned out, Bohr's philosophical specificity and radicality reside precisely in this dynamic conception. For this reason, as will be summarized below, it is also in the dynamic conception that complementarity has the most important intersections with hermeneutic philosophy and, in a different way, with Derridean deconstruction.

(4) As we saw in Chapter 5, Bohr's idea of complementarity as the relation between the roles of 'actors' and 'spectators' is generally parallel to the hermeneutic notion of the relation between "belonging" and "distanciation" or "alienation" as it is thematized by philosophers such as Gadamer and Ricoeur. In questioning the notion of the pure 'spectator' and emphasizing the irreducible moment of the 'actor,' Bohr's approach is similar to Gadamer's as well as Ricoeur's with their appeal to the irreducible moment of belonging.

I have further specified this philosophical affinity in terms of Bohr's different conceptions of complementarity. (a) As regards his 'early' static-contrastive conception, the above analogy is considerably limited. For, in this conception, the roles of 'actors' and 'spectators' simply stand side by side, while Gadamer's as well as Ricoeur's philosophy concerns the process of transition from belonging to alienation/distanciation and vice versa. (b) As for Bohr's later static-symmetrical conception, which juxtaposes the complementary relata as similar items, we can no longer speak of any meaningful intersection with hermeneutic philosophy. (c) In contrast, his dynamic conception of complementarity – with regard to non-physical fields as well as to quantum theory as reinterpreted in the 'middle' period – is found to be in close proximity specifically to Ricoeur's idea of the dialectic of event and meaning as part of the dialectic of belonging and distanciation. Even here, however, there is a subtle difference between the two thinkers' approaches. That is, in Bohr's dynamic conception, the 'spectator's' reflection on the 'actor's' involvement proves to have itself a character of involvement, whereas, in Ricoeur's thought, belonging and distanciation are each determined as such and are put in order as definite stages of dialectics.

One of the other major differences between Bohr's complementarity and hermeneutic philosophy is that the former does not privilege the 'actor's' involvement over the 'spectator's' detachment, while Gadamer and Ricoeur conceive belonging as primary and original as against alienation or distanciation. We can thus see that, despite the overall similarity between Bohr's complementarity and hermeneutic philosophy, they differ from each other in some important respects.

(5) Finally, in Chapter 6, I examined the conceptual intersections between Bohr's complementarity and Derridean deconstruction. Bohr's overall approach may be characterized as a project of questioning and dislodging the privilege hitherto accorded to the 'spectator's' detachment over the 'actor's' involvement, and this constitutes its basic point of contact with Derridean deconstruction, which seeks to dismantle the system of hierarchical binary oppositions in the Western philosophical tradition.

On closer inspection, however, this linkage between the two thinkers' ideas is further specified and delimited. Among Bohr's different conceptions of complementarity, (a) the static-contrastive conception in the 'early' period, which is directed against the hierarchy of 'spectator/actor,' but not against the binary framework itself, may only in a limited way be compared with Derridean deconstruction. (b) His later static-symmetrical conception, which is devoid of the specific direction in which the hierarchical order is disorganized, is quite foreign to Derrida's thought. (c) It is thus solely the dynamic conception that bears an essential affinity with Derridean deconstruction, an affinity in a sense even closer than with hermeneutic philosophy. That is, Bohr's dynamic conception has in common with Derridean deconstruction – specifically its critique of the "transcendental signified" – the crucial idea that any access to the signified is subject to an indefinitely extensible series of signs in which the signifying structure is uncontrollably displaced. My analysis has also shown, however, at least one major contrast between the two thinkers' overall approaches. In Bohr's complementarity, the 'actor's' involvement as a moment of *proximity* serves to disrupt the traditional primacy of the 'spectator's' detachment as a moment of *distance*, whereas Derridean deconstruction proceeds by virtue of 'distancing' moments such as *différance* to dismantle the primacy of presence as absolute proximity. In this way, the general strategic roles of proximity and distance in Bohr's and Derrida's projects are inverse to each other.

In the context of this Bohr–Derrida relation, I have also assessed Plotnitsky's analysis of the subject, an innovative rereading of Bohr which nevertheless has certain interpretive problems. While making important points of analogy such as that between Bohr's displaced notion of reality and Derrida's concept of "radical alterity," between Bohr's "complementarization" of conceptual pairs and Derrida's undoing of hierarchical binary oppositions, and so on, Plotnitsky fails to focus on the specific *directionality* in which Bohr transforms the traditional hierarchy – a crucial point in my interpretation at which the two thinkers' approaches are both closely linked and differentiated.

On the whole, the present study as summarized so far has indicated how the philosophical implications of Bohr's complementarity extend not only beyond the confines of quantum physics, but also beyond the conventional frames of the philosophy of science. What I have called the dynamic conception of complementarity, in particular, has been shown to exceed the narrow interpretive schemes of many prior studies, and instead to require a new interpretive approach open to a wide range of philosophical orientations including hermeneutics and deconstruction. To conclude the book, I wish to make a few final remarks on some further implications of this study that I have so far only indirectly hinted at. As we have seen, particularly in Chapter 4, Bohr's point of view is such that complementarity in quantum theory (or any other field of science) and complementarity in epistemology do not simply stand side by side, but rather are internally connected, since quantum theory is one of the fields in which we use and define concepts and words while at the same time materially engaging with nature. This implies that our material intervention in natural processes and our conceptual thinking and discourse on such intervention no longer belong to separate areas, but are jointly situated in the framework of complementarity. Specifically in the dynamic conception of complementarity, any discourse, even of a detached descriptive type, on our practical involvement in natural phenomena turns out to have itself a character of involvement in such a way that it uncontrollably alters the signifying structure comprising both discursive and material elements. In the modern philosophical tradition, while the Cartesian mind-body dualism as such has been considerably discredited, the conceptual-discursive and the material-physical – and, by extension, the cultural and the natural – have still largely been treated as disjunctive and functionally autonomous spheres. In recent years, however, precisely this dichotomy of material/discursive or natural/cultural has been increasingly questioned and contested in some important philosophical and sociological approaches to science.[1] The present work suggests the possibility that an analysis of Bohr's complementarity sheds new light on this pivotal problematic of contemporary science studies, not by simply dissolving the discursive/material distinction, but by seeking to reinscribe it in a non-dichotomous, non-metaphysical conception of culture and nature, the human and the nonhuman, and so forth.

Further, in this connection, let us again recall the thematic of conceptual containment in my reading of Bohr. What I have characterized as the impossibility of full conceptual containment is without doubt closely linked to Bohr's explicit notion of uncontrollable object–instrument interaction in quantum theory. Given the above inseparability of the conceptual-discursive and the material-physical, the mechanism of conceptual containment cannot be dissociated from the attempt at technical control of phenomena, and therefore – and still more importantly – the dynamic

[1] Relevant works along these lines include Latour (1991), Rouse (2002), and Barad (2007). This naturally does not mean, however, that the point of view of the present study is in full accord with any of the works just cited.

complementary relation between conceptual containment and its disruption is bound up with the same kind of relation between technical control and its breakdown. In the context of contemporary science studies, with increasing emphasis laid on the material and technological dimension of science, our consideration thus suggests an approach to the technical controllability *and* uncontrollability of nature as coupled with the question of the possibility and impossibility of conceptual-discursive containment – which in turn may further be connected with the thematic of radical alterity as discussed in the final chapter. In this way, the present study of Bohr's complementarity can hopefully help open up a new philosophical perspective on science, nature, culture, and other related categories, with a focus upon, among other issues, the possibility and impossibility of our cognitive as well as technical control of both the world and ourselves.

References

In the citations, either the abbreviated title or the author(s) and year of publication are followed by the page number(s). In the case of books in series, the volume number precedes the page number with a colon in between (e.g. 2:168). Where two page numbers separated by a slash are given (e.g. 48/63), the first number refers to the text in its original language, and the second to the English translation, unless otherwise indicated.

I use only the following two abbreviated titles for works by Niels Bohr:

NBCW *Niels Bohr Collected Works*, 12 vols. Amsterdam: North-Holland Publishing Company, 1972–2007.

PWNB *The Philosophical Writings of Niels Bohr*, 4 vols. Woodbridge, Conn.: Ox Bow Press, 1987–1998.

The other works are cited by reference to the authors and years of publication as given in the list below. The authors' names are largely omitted, however, when they are obvious from the context. For the works for which both the original text and the English translation are indicated in the list, I cite the year of publication only of the original (e.g. Gadamer 1972, 266/282).

Apel, Karl-Otto. 1973. *Transformation der Philosophie*, Vol. 1: *Sprachanalytik, Semiotik, Hermeneutik*; Vol. 2: *Das Apriori der Kommunikationsgemeinschaft*. Frankfurt am Main: Suhrkamp. Partial Eng. trans. Glyn Adey and David Frisby. *Towards a Transformation of Philosophy*. London: Routledge & Kegan Paul, 1980.

Barad, Karen. 1996. "Meeting the Universe Halfway: Realism and Social Constructivism Without Contradiction." In *Feminism, Science, and the Philosophy of Science*, edited by Lynn Hankinson Nelson and Jack Nelson, 161–194. Dordrecht: Kluwer Academic Publishers.

Barad, Karen. 1998. "Getting Real: Technoscientific Practices and the Materialization of Reality." *Differences* 10 (2): 87–126.

Barad, Karen. 2007. *Meeting the Universe Halfway: Quantum Physics and the Entanglement of Matter and Meaning*. Durham: Duke University Press.

Bataille, George. 1967. *La part maudite,* précédé de *La notion de dépense*. Paris: Les Éditions de Minuit. Eng. trans. Robert Hurley. *The Accursed Share*, Vol. 1. New York: Zone Books, 1991.

Beller, Mara. 1999. *Quantum Dialogue: The Making of a Revolution*. Chicago: The University of Chicago Press.

Beller, Mara, and Arthur Fine. 1994. "Bohr's Response to EPR." In Faye and Folse 1994, 1–31.

Bertram, Georg W. 2002. *Hermeneutik und Dekonstruktion*. München: Wilhelm Fink Verlag.

Bitbol, Michel. 1996. *Schrödinger's Philosophy of Quantum Mechanics*. Vol. 188 of *Boston Studies in the Philosophy of Science*. Dordrecht: Kluwer Academic Publishers.

Bohm, David. 1957. *Causality and Chance in Modern Physics*. London: Routledge and Kegan Paul, 1984.

Bohr, Niels. 1985. *Niels Bohr 1885–1962: Der Kopenhagener Geist in der Physik*, edited by Karl v. Meyenn, Klaus Stolzenburg, and Roman U. Sexl. Braunschweig: Friedf. Vieweg & Sohn.

Bohr, Niels. 1991. *Physique quantique et connaissance humaine*, Translated by Edmond Bauer and Roland Omnès. Paris: Éditions Gallimard.

Brock, Steen. 2003. *Niels Bohr's Philosophy of Quantum Physics in the Light of the Helmholtzian Tradition of Theoretical Physics*. Berlin: Logos Verlag.

Bunge, Mario. 1955. "Strife About Complementarity." *The British Journal for the Philosophy of Science* 6 (21): 1–12.

Bunge, Mario, ed. 1967. *Quantum Theory and Reality*. Berlin: Springer.

Bunge, Mario. 1988. "Niels Bohr's Philosophy." *Philosophia Naturalis* 25: 399–415.

Camilleri, Kristian. 2009. *Heisenberg and the Interpretation of Quantum Mechanics: The Physicist as Philosopher*. Cambridge: Cambridge University Press.

Caputo, John D. 1987. *Radical Hermeneutics: Repetition, Deconstruction, and the Hermeneutic Project*. Bloomington: Indiana University Press.

Cassidy, David C. 1992. *Uncertainty: The Life and Science of Werner Heisenberg*. New York: W. H. Freeman and Company.

Cassirer, Ernst. 1977. *Zur modernen Physik*. Darmstadt: Wissenschaftliche Buchgesellschaft.

Chevalley, Catherine. 1994. "Niels Bohr's Words and the Atlantis of Kantianism." In Faye and Folse 1994, 33–55.

Chevalley, Catherine. 1995. "Philosophy and the Birth of Quantum Theory." In: *Physics, Philosophy and the Scientific Community*. Vol. 163 of *Boston Studies in the Philosophy of Science*, edited by Kostas Gavroglu et al., 11–37. Dordrecht: Kluwer Academic Publishers.

Critchley, Simon. 1992. *The Ethics of Deconstruction: Derrida and Levinas*. Oxford: Blackwell Publishers.

Culler, Jonathan. 1982. *On Deconstruction: Theory and Criticism After Structuralism*. Ithaca, NY: Cornell University Press.

Darrigol, Olivier. 1992. *From c-Numbers to q-Numbers: Classical Analogy in the History of Quantum Theory*. Berkeley: University of California Press.

Derrida, Jacques. 1962. "Introduction" to Edmund Husserl's *L'origine de la géométrie*, 3–171. Paris: Presses Universitaires de France. Eng. trans. John P. Leavy, Jr. *Edmund Husserl's Origin of Geometry: An Introduction*, 23–153. New York: Nicolas Hays, 1978.

Derrida, Jacques. 1967a. *L'écriture et la différence*. Paris: Éditions du Seuil. Eng. trans. Alan Bass. *Writing and Difference*. Chicago: The University of Chicago Press, 1978.

Derrida, Jacques. 1967b. *De la grammatologie*. Paris: Les Éditions de Minuit. Eng. trans. Gayatri Chakravorty Spivak. *Of Grammatology*. Baltimore: Johns Hopkins University Press, 1976.

Derrida, Jacques. 1967c. *La voix et le phénomène*. Paris: Presses Universitaires de France. Eng. trans. David B. Allison. *Speech and Phenomena and Other Essays on Husserl's Theory of Signs*. Evanston, IL: Northwestern University Press, 1973.

Derrida, Jacques. 1972a. *La dissémination*. Paris: Éditions du Seuil. Eng. trans. Barbara Johnson. *Dissemination*. Chicago: The University of Chicago Press, 1981.

Derrida, Jacques. 1972b. *Marges de la philosophie*. Paris: Les Éditions de Minuit. Eng. trans. Alan Bass. *Margins of Philosophy*. Chicago: The University of Chicago Press, 1982.

Derrida, Jacques. 1972c. *Positions*. Paris: Les Éditions de Minuit. Eng. trans. Alan Bass. *Positions*. Chicago: The University of Chicago Press, 1987.

Derrida, Jacques. 1990. *Limited Inc*. Paris: Éditions Galilée. Eng. trans. Samuel Weber and Jeffrey Mehlman. *Limited Inc*. Evanston, IL: Northwestern University Press, 1988.

Derrida, Jacques. 1994. *Force de loi: Le «Fondement mystique de l'autorité»*. Paris: Éditions Galilée. Eng. trans. Mary Quaintance. "The Force of Law: 'The Mystical Foundation of Authority.'" In *Deconstruction and the Possibility of Justice*, edited by Drucilla Cornell et al., 68–91. New York: Routledge, 1992.

Derrida, Jacques. 2001. *Papier machine: Le ruban de machine à écrire et autres réponses*. Paris: Éditions Galilée. Eng. trans. Rachel Bowlby. *Paper Machine*. Stanford: Stanford Univeristy Press, 2005.

Derrida, Jacques, and Hans-Georg Gadamer. 2004. *Der ununterbrochene Dialog*. Frankfurt am Main: Suhrkamp Verlag.

Dostal, Robert J., ed. 2002. *The Cambridge Companion to Gadamer*. Cambridge: Cambridge University Press.

Einstein, Albert. 1905. "Über einen die Erzeugung und Verwandlung des Lichtes betreffenden heuristischen Gesichtspunkt." *Annalen der Physik* 17: 132–48.

Einstein, Albert. 1979. *Aus meinen späten Jahren*. Stuttgart: Deutsche Verlags-Anstalt. Eng. trans. *Out of My Later Years*. New York: Philosophical Library, 1950.

Einstein, Albert, Boris Podolsky, and Nathan Rosen. 1935. "Can Quantum-Mechanical Description of Physical Reality Be Considered Complete?" *Physical Review* 47: 777–80.

Evans, J. Claude. 1991. *Strategies of Deconstruction: Derrida and the Myth of the Voice*. Minneapolis: University of Minnesota Press.

Falkenburg, Brigitte. 1998. "Bohr's Principles of Unifying Quantum Disunities." *Philosophia Naturalis* 35: 95–120.

Favrholdt, David. 1992. *Niels Bohr's Philosophical Background*. Copenhagen: Det Kongelige Danske Videnskabernes Selskab.

Faye, Jan. 1991. *Niels Bohr: His Heritage and Legacy — An Anti-Realist View of Quantum Mechanics*. Dordrecht: Kluwer Academic Publishers.

Faye, Jan. 2002a. "Copenhagen Interpretation of Quantum Mechanics." In *The Stanford Encyclopedia of Philosophy (Summer 2002 Edition)*, edited by Edward N. Zalta. http://plato.stanford.edu/archives/sum2002/entries/qm-copenhagen/

Faye, Jan. 2002b. *Rethinking Science: A Philosophical Introduction to the Unity of Science*. Aldershot: Ashgate Publishing Limited.

Faye, Jan, and Henry J. Folse, eds. 1994. *Niels Bohr and Contemporary Philosophy*. Vol. 153 of *Boston Studies in the Philosophy of Science*. Dordrecht: Kluwer Academic Publishers.

Faye, Jan, and Henry J. Folse. 1998. "Introduction." In PWNB, 4: 1–23.

Fehér, Márta, Olga Kiss, and László Ropolyi, eds. 1999. *Hermeneutics and Science: Proceedings of the First Conference of the International Society for Hermeneutics and Science*. Vol. 206 of *Boston Studies in the Philosophy of Science*. Dordrecht: Kluwer.

Feshbach, Herman, Tetsuo Matsui, and Alexandra Oleson, eds. 1988. *Niels Bohr: Physics and the World*. Chur: Harwood Academic Publishers.

Feyerabend, Paul K. 1981. *Scientific Papers*, Vol. 1: *Realism, Rationalism and Scientific Method*. Cambridge: Cambridge University Press.

Fine, Arthur. 1986. *The Shaky Game: Einstein, Realism, and the Quantum Theory*. Chicago: The University of Chicago Press.

Folse, Henry J. 1985. *The Philosophy of Niels Bohr: The Framework of Complementarity*. Amsterdam: North-Holland Physics Publishing.

Folse, Henry J. 1996. "The Bohr-Einstein Debate and the Philosophers' Debate over Realism Versus Anti-Realism." In *Realism and Anti-Realism in the Philosophy of Science: Beijing International Conference, 1992*, edited by Robert S. Cohen, Risto Hilpinen, and Qiu Renzong. Vol. 169 of *Boston Studies in the Philosophy of Science*, 289–98. Dordrecht: Kluwer Academic Publishers.

Forget, Philippe, ed. 1984. *Text und Interpretation*. München: Wilhelm Fink Verlag.

French, Anthony Philip, and Joseph P. Kennedy, eds. 1985. *Niels Bohr: Centenary Volume*. Cambridge, MA: Harvard University Press.

Froula, Christine. 1985. "Quantum Physics/Postmodern Metaphysics: The Nature of Jacques Derrida." *Western Humanities Review* 39: 289–313.

Fujita, Shingo. 1991. *Sōhosei no tetsugakuteki-kōsatsu* [*Philosophical Investigations on Complementarity*]. Tokyo: Taga-shuppan.

Gadamer, Hans-Georg. 1967ff. *Kleine Schriften*, 4 vols. Tübingen: J. C. B. Mohr, 1967–1977.

Gadamer, Hans-Georg. 1972. *Wahrheit und Methode: Grundzüge einer philosophischen Hermeneutik*, 3rd edn. Tübingen: J. C. B. Mohr. Eng. trans. revised by Joel Weinsheimer and Donald G. Marshall. *Truth and Method*. London: Sheed & Ward, 1989.

Gadamer, Hans-Georg. 1976. *Philosophical Hermeneutics*. Translated and edited by David E. Linge. Berkeley: University of California Press.

Gadamer, Hans-Georg. 1981. "Text and Interpretation." In Michelfelder and Palmer 1989, 21–51.

Gadamer, Hans-Georg. 1985. "*Destruktion* and deconstruction." In Michelfelder and Palmer 1989, 102–113.

Gadamer, Hans-Georg. 1986ff. *Gesammelte Werke*, 10 vols. Tübingen: J. C. B. Mohr, 1986–1999.

Gadamer, Hans-Georg. 1987. "Hermeneutics and Logocentrism." In Michelfelder and Palmer 1989, 114–125.

Gasché, Rodolphe. 1986. *The Tain of the Mirror: Derrida and the Philosophy of Reflection*. Cambridge, MA: Harvard University Press.

Gross, Paul R., and Norman Levitt. 1994. *Higher Superstition: The Academic Left and Its Quarrels with Science*. Baltimore: The Johns Hopkins University Press.

Hahn, Lewis Edwin, ed. 1995. *The Philosophy of Paul Ricoeur* [*The Library of Living Philosophers*, Vol. XXII]. Chicago: Open Court.

Heelan, Patrick A. 1965. *Quantum Mechanics and Objectivity: A Study of the Physical Philosophy of Werner Heisenberg*. The Hague: Martinus Nijhoff.

Heelan, Patrick A. 1970. "Complementarity, Context Dependence, and Quantum Logic." *Foundations of Physics* 1 (2): 95–110.

Heelan, Patrick A. 1977. "Hermeneutics of Experimental Science in the Context of Life-World." In *Interdisciplinary Phenomenology*, edited by Don Ihde and Richard M. Zaner, 7–50. The Hague: Martinus Nijhoff.

Heelan, Patrick A. 1983. *Space-Perception and the Philosophy of Science*. Berkerley: University of California Press.

Hegel, Georg Wilhelm Friedrich. 1807. *Phänomenologie des Geistes*. Hamburg: Felix Meiner Verlag, 1952. Eng. trans. A. V. Miller. *Phenomenology of Spirit*. Oxford: Oxford University Press, 1977.

Heidegger, Martin. 1927. *Sein und Zeit*. Tübingen: Max Niemeyer Verlag, 1986. Eng. trans. John Macquarrie and Edward Robinson. *Being and Time*. Oxford: Basil Blackwell, 1962.

Heisenberg, Werner. 1930. *Physikalische Prinzipien der Quantentheorie*. Stuttgart: S. Hirzel Verlag, 1958. Eng. trans. Carl Eckart and F. C. Hoyt. *The Physical Principles of the Quantum Theory*. New York: Dover Publications, 1949.

Heisenberg, Werner. 1958. *Physics and Philosophy: The Revolution in Modern Science*. New York: Harper & Row, Publishers.

Heisenberg, Werner. 1967. "Quantum Theory and Its Interpretation." In Rozental 1967, 94–108.

Heisenberg, Werner. 1969. *Der Teil und das Ganze: Gespräch im Umkreis der Atomphysik*. München: R. Piper & Co. Verlag. Eng. trans. Arnold J. Pomerans. *Physics and Beyond: Encounters and Conversations*. New York: Harper & Row, Publishers, 1971.

Heisenberg, Werner. 1971. *Schritte über Grenzen*. München: R. Piper & Verlag.

Heisenberg, Werner. 1984ff. *Werner Heisenberg Gesammelte Werke/ Collected Works*, A1-3. Berlin: Springer Verlag, 1984–1993.

Held, Carlsten. 1994. "The Meaning of Complementarity." *Studies in History and Philosophy of Science* 25 (6): 871–93.

Held, Carlsten. 1999. *Die Bohr-Einstein-Debatte: Quantenmechanik und physikalische Wirklichkeit*. Paderborn: Mentis Verlag.

Hendry, John. 1984. *The Creation of Quantum Mechanics and the Bohr-Pauli Dialogue*. Dordrecht: D. Reidel Publishing Company.

Holton, Gerald. 1973. *Thematic Origins of Scientific Thought: Kepler to Einstein*. Cambridge, MA: Harvard University Press.

Honner, John. 1987. *The Description of Nature: Niels Bohr and the Philosophy of Quantum Physics*. Oxford: Clarendon Press.

Hooker, Clifford A. 1994. "Bohr and the Crisis of Empirical Intelligibility: An Essay on the Depth of Bohr's Thought and Our Philosophical Ignorance." In Faye and Folse 1994, 155–99.

Husserl, Edmund. 1913. *Logische Untersuchungen*, Vol. 2: *Untersuchungen zur Phänomenologie und Theorie der Erkenntnis*, Part 1. Tübingen: Max Niemeyer Verlag.

Ihde, Don. 1991. *Instrumental Realism: The Interface Between Philosophy of Science and Philosophy of Technology*. Bloomington: Indiana University Press.

Ihde, Don. 1998. *Expanding Hermeneutics: Visualism in Science*. Evanston, IL: Northwestern University Press.

Ito, Kenji. 2002. "Values of 'Pure Science': Nishina Yoshio's wartime discourse between nationalism and physics, 1940–1945." *Historical Studies of the Physical Sciences* 33, Part 1:61–86.

Jacobsen, Anja Skaar. 2007. "Leon Rosenfeld's Marxist Defense of Complementarity." *Historical Studies in the Physical and Biological Sciences* 37, Supplement: 3–34.

Jammer, Max. 1974. *The Philosophy of Quantum Mechanics: Interpretations of Quantum Mechanics in Historical Perspective*. New York: A Wiley-Interscience Publication.

Jammer, Max. 1989. *The Conceptual Development of Quantum Mechanics*, 2nd edn. [*The History of Modern Physics 1800–1950*, Volume 12]. New York: American Institute of Physics.

Johnson, Patricia Altenbernd. 2000. *On Gadamer*. Belmont: Wadsworth.

Kaiser, David. 1992. "More Roots of Complementarity: Kantian Aspects and Influences." *Studies in History and Philosophy of Science* 23 (2): 213–39.

Kanamori, Osamu. 2000. *Science Wars*. Tokyo: Tokyo-daigaku-shuppankai.

Kant, Immanuel. 1781. *Kritik der reinen Vernunft*. Hamburg: Felix Meiner Verlag, 1956. Eng. trans. Norman Kemp Smith. *Critique of Pure Reason*. New York: St Martin's Press, 1929.

Kant, Immanuel. 1790. *Kritik der Urteilskraft*. Hamburg: Felix Meiner Verlag, 1924.

Katsumori, Makoto. 1992. "The Theories of Relativity and Einstein's Philosophical Turn." *Studies in History and Philosophy of Science* 23 (4): 557–92.

Katsumori, Makoto. 2010. "Derridean Deconstruction and the Question of Nature." *Derrida Today* 3 (1): 56–74.

Komesaroff, Paul A. 1986. *Objectivity, Science and Society: Interpreting Nature and Society in the Age of the Crisis of Science*. London: Routledge & Kegan Paul.

Kuhn, Thomas S. 1978. *Black-Body Theory and the Quantum Discontinuity 1894–1912*. Oxford: Oxford University Press.

Latour, Bruno. 1991. *Nous n'avons jamais été modernes: Essai d'anthropologie symétrique*. Paris: Éditions La Découverte. Eng. trans. Catherine Porter. *We Have Never Been Modern*. New York: Harvester Wheatsheaf, 1993.

Lawlor, Leonard. 1992. *Imagination and Chance: The Difference Between the Thought of Ricoeur and Derrida*. Albany: State University of New York Press.

Llewelyn, John. 1986. *Derrida on the Threshold of Sense*. Houndmills: Macmillan Press.

Lyotard, Jean-François. 1979. *La condition postmoderne: Rapport sur le savoir*. Paris: Les Éditions de Minuit. Eng. trans. Geoff Bennington and Brian Massumi. *The Postmodern Condition: A Report on Knowledge*. Minneapolis: University of Minnesota Press, 1984.

MacKinnon, Edward M. 1982. *Scientific Explanation and Atomic Physics*. Chicago: The University of Chicago Press.

MacKinnon, Edward M. 1985. "Bohr on the Foundations of Quantum Theory." In French and Kennedy 1985, 101–20.

Makita, Etsurō. 1997. *Ricoeur no text-kaishakugaku* [*Ricoeur's Hermeneutics of the Text*]. Kyoto: Kōyō-shobō.

Meyer-Abich, Klaus Michael. 1965. *Korrespondenz, Individualität und Komplementarität: Eine Studie zur Geistesgeschichte der Quantentheorie in den Beiträgen Niels Bohrs*. Wiesbaden: Franz Steiner Verlag.

Michelfelder, Diane P., and Richard E. Palmer, eds. 1989. *Dialogue and Deconstruction: The Gadamer-Derrida Encounter*. Albany: State University of New York Press.

Mueller-Vollmer, Kurt. 1985. "Introduction: Language, Mind, and Artefact: An Outline of Hermeneutic Theory Since the Enlightenment." In *The Hermeneutics Reader*, edited by Kurt Mueller-Vollmer, 1–53. New York: The Continuum Publishing Co.

Murdoch, Dugald. 1987. *Niels Bohr's Philosophy of Physics*. Cambridge: Cambridge University Press.

Pais, Abraham. 1991. *Niels Bohr's Times: In Physics, Philosophy, and Polity*. Oxford: Clarendon Press.

Pauli, Wolfgang. 1961. *Physik und Erkenntnistheorie*. Braunschweig: Friedf. Vieweg & Sohn, 1984.

Petersen, Aage. 1963. "The Philosophy of Niels Bohr." In French and Kennedy 1985, 299–310.

Petruccioli, Sandro. 1993. *Atoms, Metaphors and Paradoxes: Niels Bohr and the Construction of a New Physics*, Translated by Ian McGilvray. Cambridge: Cambridge University Press.

Pitt, Axel. 1971. "Die dialektische Begründung der quantenmechanischen Statistik durch die Metaphysik Hegels." *Philosophia Naturalis* 13: 371–93.

Planck, Max. 1901. "Über das Gesetz der Energieverteilung im Normalspektrum." *Annalen der Physik* 4 (4): 553–63.

Plotnitsky, Arkady. 1993a. *Reconfigurations: Critical Theory and General Economy*. Gainesville: University Press of Florida.

Plotnitsky, Arkady. 1993b. *In the Shadow of Hegel: Complementarity, History, and the Unconscious*. Gainesville: University Press of Florida.

Plotnitsky, Arkady. 1994. *Complementarity: Anti-Epistemology After Bohr and Derrida*. Durham: Duke University Press.

Plotnitsky, Arkady. 1997. "'But It Is Above All Not True': Derrida, Relativity, and the 'Science Wars.'" *Postmodern Culture* 7 (2). http://muse.jhu.edu/journals/postmodern_culture/v007/7.2plotnitsky.html

Plotnitsky, Arkady. 2001a. "Algebra and Allegory: Nonclassical Epistemology, Quantum Theory and the Work of Paul de Man." In *Material Events: Paul de Man and the Afterlife of Theory*, edited by Tom Cohen et al., 49–89. Minneapolis: The University of Minneapolis Press.

Plotnitsky, Arkady. 2001b. "Reading Bohr: Complementarity, Epistemology, Entanglement, and Decoherence." In *Decoherence and Its Implications in Quantum Computation and Information Transfer*, edited by Tony Gonis and Patrice E. A. Turchi, 3–37. Amsterdam: IOS Press.

Plotnitsky, Arkady. 2002. *The Knowable and the Unknowable: Modern Science, Nonclassical Thought, and the "Two Cultures."* Ann Arbor: The University of Michigan Press.

Plotnitsky, Arkady. 2003. "Mysteries Without Mysticism and Correlations Without Correlata: On Quantum Knowledge and Knowledge in General." *Foundations of Physics* 33 (11): 1649–689.

Plotnitsky, Arkady. 2006. *Reading Bohr: Physics and Philosophy*. Dordrecht: Springer.

Popper, Karl R. 1982. *Quantum Theory and the Schism in Physics* [From the Postscript to *The Logic of Scientific Discovery*], edited by William Warren Bartley, III. London: Hutchinson & Co.

Radder, Hans. 1988. *The Material Realization of Science: A Philosophical View on the Experimental Natural Sciences, Developed in Discussion with Habermas*, Translated by Tineke van Putten and Dawn Wolfswinkel. Assen: Van Gorcum & Comp. Also digitally available at http://dare.ubvu.vu.nl/handle/1871/13151

Radder, Hans. 1996. *In and About the World: Philosophical Studies of Science and Technology*. Albany: State University of New York Press.

Radder, Hans. 2002. "How Concepts Both Structure the World and Abstract from It." *The Review of Metaphysics* 55: 581–613.

Radder, Hans. 2006. *The World Observed/The World Conceived*. Pittsburgh, PA: The University of Pittsburgh Press.

Ricoeur, Paul. 1969. *Le conflit des interprétations: Essais d'herméneutique*. Paris: Éditions du Seuil. Ed. Don Ihde, Eng. trans. Kathleen McLaughlin et al. *The Conflict of Interpretations: Essays in Hermeneutics*. Evanston, IL: Northwestern University Press, 1974.

Ricoeur, Paul. 1971. "Événement et sens dans le discours." In *Paul Ricoeur: ou la liberté selon l'espérance*, edited by Michel Philibert, 177–87. Paris: Éditions Seghers.

Ricoeur, Paul. 1975. *La métaphore vive*. Paris: Éditions du Seuil. Eng. trans. Robert Czerny. *The Rule of Metaphor: Multi-Disciplinary Studies of the Creation of Meaning in Language*. Toronto: University of Toronto Press, 1977.

Ricoeur, Paul. 1976. *Interpretation Theory: Discourse and the Surplus of Meaning*. Fort Worth: Texas Christian University Press.

Ricoeur, Paul. 1981. *Hermeneutics and the Human Sciences*, edited and translated by John B. Thompson. Cambridge: Cambridge University Press.

Ricoeur, Paul. 1986. *Du texte à l'action: Essais d'herméneutique, II*. Paris: Éditions du Seuil. Eng. trans. Kathleen Blamey and John B. Thompson. *From Text to Action: Essays in Hermeneutics, II*. Evanston, IL: Northwestern University Press, 1991.

Ricoeur, Paul. 1995. "Intellectual Autobiography," Translated by Kathleen Blamey. In Hahn 1995, 3–53.

Röseberg, Ulrich. 1995. "Did They Just Misunderstand Each Other? Logical Empiricists and Bohr's Complementarity Argument." In *Physics, Philosophy and the Scientific Community*, edited by Gavroglu et al. Vol. 163 of *Boston Studies in the Philosophy of Science*, 105–23. Dordrecht: Kluwer Academic Publishers.

Rosenfeld, Léon. 1967. "Niels Bohr in the Thirties: Consolidation and Extension of the Conception of Complementarity." In Rozental 1967, 114–36.

Ross, Andrew, ed. 1996. *Science Wars*. Durham: Duke University Press.

Rouse, Joseph. 1987. *Knowledge and Power: Toward a Political Philosophy of Science*. Ithaca: Cornell University Press.

Rouse, Joseph. 2002. *How Scientific Practices Matter: Reclaiming Philosophical Naturalism*. Chicago: The University of Chicago Press.

Rozental, Stefan, ed. 1967. *Niels Bohr: His Life and Work as Seen by His Friends and Colleagues*. Amsterdam: North-Holland Physics Publishing.

Saussure, Ferdinand de. 1916. *Cours de linguistique générale*. Édition critique préparée par Tullio de Mauro. Paris: Editions Payot, 1972. Eng. trans. Roy Harris. *Course in General Linguistics*. La Salle, IL: Open Court Publishing Company, 1983.

Schiemann, Gregor. 2008. *Werner Heisenberg*. Nördlingen: Verlag C. H. Beck.

Schrödinger, Erwin. 1926. "Quantisierung als Eigenwertproblem I-IV." *Annalen der Physik* 79 (1926): 361–76, 489–527; 80 (1926): 437–90; 81 (1926): 109–39.

Selleri, Franco. 1990. *Die Debatte um die Quantentheorie*. Braunschweig: Verlag Vieweg.

Smith, Barbara Herrnstein, and Arkady Plotnitsky. 1997. "Introduction: Networks and Symmetries, Decidable and Undecidable." In *Mathematics, Science, and Postclassical Theory*, edited by Smith, Barbara Herrnstein, and Arkady Plotnitsky, 1–16. Durham: Duke University Press.

Sokal, Alan. 1996a. "Transgressing the Boundaries: Toward a Transformative Hermeneutics of Quantum Gravity." *Social Text* 46/47:217–52. Reprinted in Sokal and Bricmont, *Intellectual Impostures: Postmodern Philosophers' Abuse of Science*, 199–240. London: Profile Books, 1998.

Sokal, Alan. 1996b. "Transgressing the Boundaries: An Afterword." *Dissent* 43 (4): 93–9. Reprinted in Sokal and Bricmont, *Intellectual Impostures: Postmodern Philosophers' Abuse of Science*, 248–58. London: Profile Books, 1998..

Sokal, Alan, and Jean Bricmont. 1997. *Impostures Intellectuelles*. Paris: Editions Odile Jacob. Eng. trans. *Intellectual Impostures: Postmodern Philosophers' Abuse of Science*. London: Profile Books, 1998.

Takahashi, Tetsuya. 1992. *Gyakkō no logos* [*Logos in Backlight*]. Tokyo: Miraisha.

Takahashi, Tetsuya. 1998. *Derrida: Datsukōchiku* [*Derrida: Deconstruction*]. Tokyo: Kōdansha.

Thompson, John B. 1981. *Critical Hermeneutics: A Study in the Thought of Paul Recoeur and Jürgen Habermas*. Cambridge: Cambridge University Press.

van der Waerden, Bartel Leendert, ed. 1967. *Sources of Quantum Mechanics*. Amsterdam: North-Holland Publishing Co.

von Weizsäcker, Carl Friedrich. 1971. *Die Einheit der Natur*. München: Carl Hanser Verlag. Eng. trans. Francis J. Zucker. *The Unity of Nature*. New York: Farrar, Straus and Giroux, 1980.

von Weizsäcker, Carl Friedrich. 1976. *Zum Weltbild der Physik*. Stuttgart: S. Hirzel Verlag.

von Weizsäcker, Carl Friedrich. 1983. *Wahrnehmung der Neuzeit*. München: Carl Hanser Verlag.

Wheeler, John Archibald, and Wojciech Hubert Zurek, eds. 1983. *Quantum Theory and Measurement*. Princeton: Princeton University Press.

Wheeler III, Samuel C. 2000. *Deconstruction as Analytic Philosophy*. Stanford: Stanford University Press.

Wittgenstein, Ludwig. 1958. *Philosophische Untersuchungen*. Oxford: Basil Blackwell. Eng. trans. G. E. M. Anscombe. *Philosophical Investigations*. Oxford: Basil Blackwell, 1958.

Žižek, Slavoy. 1996. *The Indivisible Remainder: An Essay on Schelling and Related Matters*. London: Verso.

Name Index

Note: The letter **n** following the locators refers to **notes** cited in the text.

Subject Index

Note: The letter **n** following the locators refers to **notes** cited in the text.

W

CPSIA information can be obtained at www.ICGtesting.com
Printed in the USA
LVOW01*1351031113

359804LV00003B/392/P

9 789400 717473